American Institute of Chemical Engineers

Process Development Division Conference

Presentations at the
2008 AIChE Spring National Meeting

April 6-10, 2008
New Orleans, Louisiana, USA

Printed from e-media with permission by:

Curran Associates, Inc.
57 Morehouse Lane
Red Hook, NY 12571
www.proceedings.com

ISBN: 978-1-60560-220-2

Some format issues inherent in the e-media version may also appear in this print version.

ISBN: 978-1-60560-220-2

American Institute of Chemical Engineers
Proceedings
Three Park Avenue
New York, NY 10016-5991
Phone: 212-591-8100

www.aiche.org

American Institute of Chemical Engineers

Process Development Division Conference
2008

TABLE OF CONTENTS

AIChE National Meeting:
ENHANCING PLANT ASSET MANAGEMENT IN PROCESS INDUSTRIES WITH WIRELESS RETROFITS

By Paul Sereiko, President & CEO, AirSprite Technologies, Inc.
Co-Chair of the ISA SP100 Marketing Working Group and member of the HART Communication Foundation

---

Wireless technology has had a significant impact on the way business is done globally. It enables a greater degree of connectivity among devices for enhanced monitoring and utilization of existing assets and has led to the development of new applications that improve productivity, uptime and overall business performance.

Yet, industrial automation is one segment of the global economy that to date has seen little impact from wireless technology. Major industries, such as oil and gas, chemical, power, and water and wastewater treatment continue to operate their plants mostly with older, hard-wired control systems. A typical process facility will have well over 1,000 measurement points, none of which currently use wireless technology, and many additional points that go unmeasured because of the cost of running wires to each. This paper discusses the need for standards-compliant, wireless, sensor-based technology in these industries for enhanced plant asset management and the benefits that will result.

Improving Plant Asset Management ~ Proactively

Most, if not all, industrial plants use networks to link devices and instruments to their control and management systems. While these systems are complex, the majority work with simple analog information, such as temperature, pressure, level and flow readings. While effective, these control and management systems can add significantly more value if they could access data that would allow them to do much more than transmit process measurements to a control system or send commands such as on/off, open/close, or setpoints that essentially tell the plant how to operate.

Many of the devices and instruments in a plant actually collect and maintain intelligent digital data about their own performance, individual processes, or the overall operation of a plant. These data can be extremely valuable: for example, it can help managers predict when a problem might occur that would force a plant shutdown. Unfortunately, most of this data is trapped in these devices with no easy way out and plant operators have a treasure trove of data that cannot be readily accessed and put to good use. Industrial managers know that productivity will improve if they have a heads up on a problem before it happens: if they can "fix it before it's broken."

One plant application area that is ripe for taking advantage of previously trapped HART data is Plant Asset Management [PAM]. The use of these PAM systems is considered a best practice for Asset Performance Management. PAM applications facilitate the management of performance, availability and reliability of plant assets by maintaining contact with all aspects of the plant, ranging from process and mechanical equipment to electrical equipment to field devices, analyzers and networks. The PAM system's responsibility is to monitor asset health, predict potential problems and failures, and optimize maintenance and operations decisions. PAM is about optimizing the performance, availability, and reliability of plant assets, which for this paper are considered to be those for machinery, production, and automation.

PAM as a practice involves:

- Monitoring asset health, in many cases this is ideally sensing in real time to detect potential problem conditions but frequently is actually on a weekly, monthly, quarterly, or even yearly basis.
- Assessing health data to predict potential problems
- Deciding on the optimal course of action for a given problem or problems
- Acting to prevent and resolve problems, such as issuing a work order to fix an instrument.

PAM application functionality usually includes:

- Commissioning management
- Calibration and compliance management
- Monitoring of smart field devices
- Analysis of field data, such as for vibration patterns or valve signatures
- Integration with Enterprise Asset Management and Computerized Maintenance Management Systems

The popularity of PAM systems is driven by several factors, including a rapidly growing number of plant assets, smaller field staffs, and the increasing rate of retirement for the aging "boomer" workforce. There are now more loops for a technician to maintain, and less expertise per technician. PAM systems are a major supplement to the workforce and enables assets to be managed effectively. Given these drivers, process manufacturing companies are realizing that maintaining competitive returns on plant assets takes more than just the manual efforts of their workforces.

According to the ARC Advisory Group, PAM systems go well beyond improving maintenance. They estimate that by employing PAM systems, the number of unplanned plant breakdowns could be reduced by nearly 45 percent, while production downtime could be cut by slightly more than 20 percent. In addition, plant managers could reduce their spare parts inventory costs by 25 percent and product defects by about 10 percent. On top of this, ARC estimates asset performance could be improved by almost 40 percent, while workforce efficiency would see about a 20 percent gain and plant availability about a 15 percent gain.

Another use of wireless access to a HART instrument is remote field device management. For instance, plant maintenance staff can cut time and effort by remotely adjusting configuration parameters, such as damping or upper and lower range settings, through their PAM system. There is no need to physically locate instruments or do handwritten reports.

Using Wireless to Retrofit Instrumentation

Major industries, such as oil and gas, chemical, power, and water and wastewater treatment continue to operate their plants mostly with older, hard-wired control systems. They have opted to upgrade plants rather than build new. In an environment where new construction isn't an economical option, plants and refineries have been retrofitting aging equipment as an alternative upgrade path to newer technology. One new industrial technology, wireless sensor networking, is emerging as a tool for retrofitting at the field device level to economically upgrade a plant for improved operational productivity. Every plant has a list of measurement points to be added to the control scheme when budget and time allow.

Using wireless technology to retrofit existing instruments by connecting directly to installed instruments allows plants to minimize downtime and production interruptions without incurring the expense of an entirely new wireless implementation. Unless new points are being added, the wiring for the old instruments is typically already in place, and the ability to take advantage of that sunk cost in ROI calculations makes it difficult to justify the premium for the entirely new wireless device and networking infrastructure.

For years, plants have been replacing older 4-20 mA field instrumentation with intelligent HART instrumentation. Intelligent devices, such as HART instruments, collect and maintain valuable digital data about their own performance, commissioning, condition, calibration, and production processes. For example, ready access to this data can help managers predict when a problem might occur that would force a plant shutdown.

With older control systems and the expense of adding HART modems to the systems, the diagnostic and digital capabilities of the HART instruments were never enabled for continuous access [figure 1]. The HART instruments were essentially used as 4-20 mA analog field instruments.

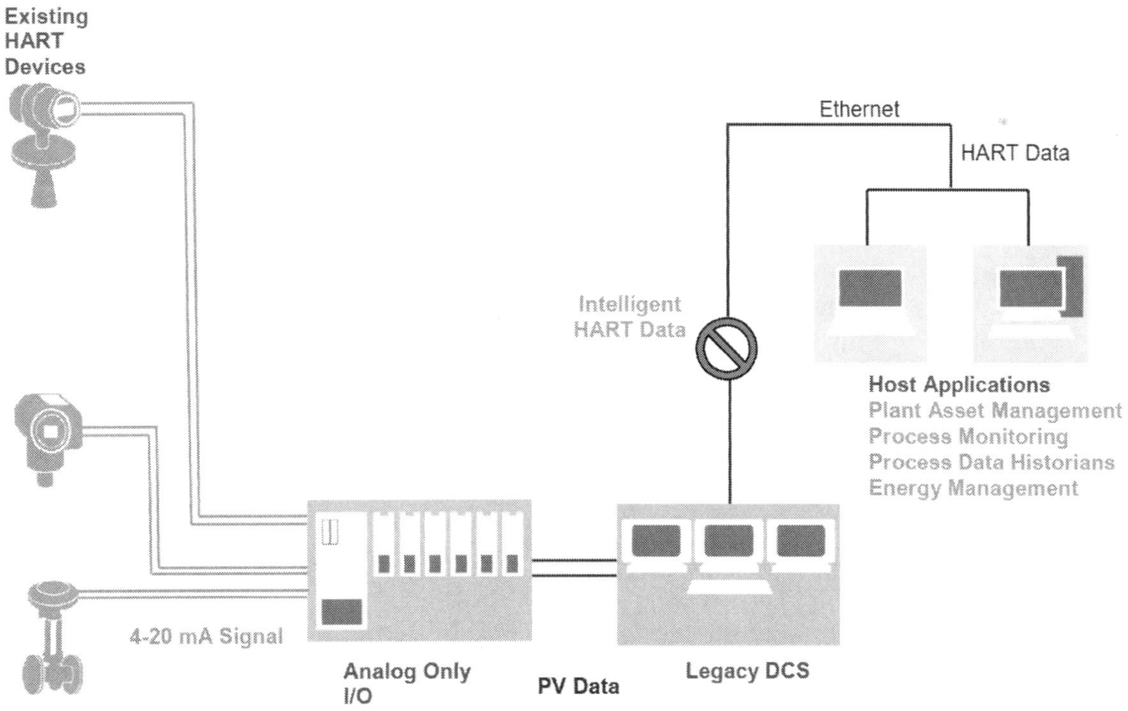

Figure 2

Aware of this problem, emerging wireless standards, such as the recently released WirelessHART and the soon to be released ISA100.11a standards, provide the capability of designing a wireless "adapter" that can be retrofitted onto existing 4-20 mA loops for the purpose of extracting digital performance data and wirelessly transmitting this information to plant applications, such as a plant asset management system or process data historian, that drive operational improvements. [Figure 2]. Automation suppliers will be introducing these adapters during 2008.

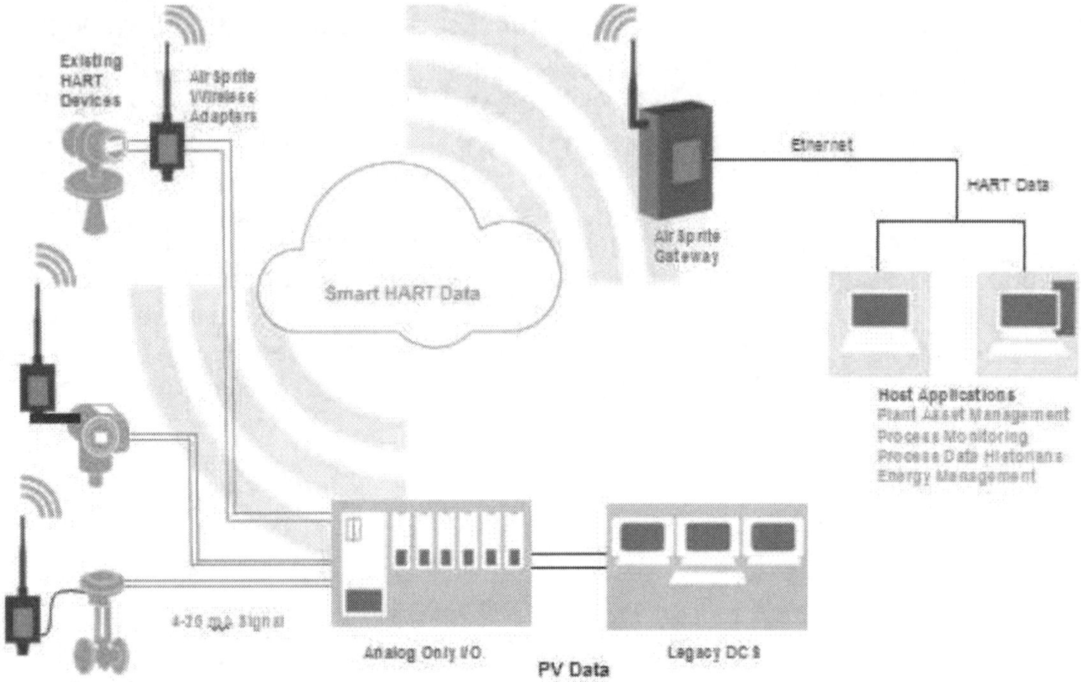

Figure 2

Beyond PAM, wireless retrofits of intelligent instruments can be valuable to control room operations. For example, digital process values can provide operators with an alternative to their analog 4-20 mA signals, especially when the analog signal has problems. These wireless retrofits can also provide completely new measurement points – if there is field power available any 4-20 mA wired instrument in combination with a wireless adapter can readily add a new measurement that is valuable for making advanced control decisions.

How the Standards are Changing the Landscape: Wireless Hart and ISA-SP100

The rapid adoption of wireless technology is being driven by the emergence of new wireless standards such as ISA100 and HART Wireless. Two standards currently under development are aimed at industrial wireless sensing solutions. The HART Communication Foundation (HCF) is working on the WirelessHART standard, which is aimed at leveraging the information collected by the nearly 25 million installed HART [Highway Addressable Remote Transducer] devices. WirelessHART will be geared specifically to the process industry, with a goal of enabling reliable, robust and secure wireless communication in real-world industrial plant applications.

The second standard is ISA100, which will support multiple protocols, including HART, as well as process and factory automation applications. The two groups are cooperating, to ensure continuity and uniformity with wireless standardization.
It is estimated that as many as 85 percent of the 25 million HART devices in use cannot directly connect their digital data to systems that manage, monitor and control industrial plants. One goal of these new standards is to enable the development of products that can unleash the power of this trapped, intelligent data, allowing easier access to information about plant assets by directly connecting many more sensors.

Today's typical wireless deployment in an industrial setting usually requires the purchase of proprietary wireless instrumentation and systems from a single vendor. The headaches accompanying this strategy include dependence on that vendor, added complexity for plant staff, and escalated project and maintenance costs. A second goal of the standards is to alleviate these headaches by allowing vendors to build infrastructure products that work with products from other vendors and with what is already installed in the plant.

So the problem is clear: Scads of incredibly valuable digital data that can improve plant efficiency and uptime and employee productivity in some of the world's most critical industries is basically sitting dormant in approximately 20 million installed devices.

A major benefit of the WirelessHART standard is that it will allow vendors to develop adapters that will be able to connect directly to installed HART devices – without changing anything on the device. These adapters will extract the intelligent HART data and then wirelessly transmit it directly to plant and enterprise applications, such as plant asset management, energy management or monitoring and control. There the data are used to do a better job of predictive maintenance, as well as to avoid major problems such as unplanned plant shutdowns. It is estimated that each HART device contains 35-40 data items that can be used to improve the performance of an industrial plant.

The Broader Landscape for Wireless Retrofits

Wireless as an enabling technology for a broad range of retrofit applications within the process industries and offers a cost effective way to modernize and realize the next great boost in productivity. Applications that transcend any specific industry segment are already being deployed. Operator mobility is enhanced with handhelds and tablet PCs, allowing operators and maintenance personal to roam their control room and shop floors. Wireless video adds process and plant security, and a host of new real time location services for plant assets and people are just around the corner.

To take advantage of these emerging applications requires a secure and robust industrial wireless infrastructure. The latest technologies and emerging standards are enabling implementations in a highly secure and robust fashion – across the enterprise. It's critical that wireless communications – like any wired networking be properly engineered, constructed, and maintained in order to perform reliably.

Once a secure and robust wireless infrastructure is constructed it must be maintained and managed to keep it functioning correctly. Wireless networks, although offering huge savings versus their wired counterparts, by their very nature, require more management. Maintaining security keys, responding to incidental or malicious interference, and managing rapidly changing technology and standards are just a few of the functions that require an expertise not necessarily available within the local IT organization.

Many organizations are finding that it's more cost effective and more secure to contract out the real time management and optimization of their wireless infrastructure. Unfortunately, many organizations simply allow their wireless networks (which will contain multiple technologies, protocols and frequencies) to grow in and ad hoc fashion. That becomes a sure way to have an unsuccessful wireless experience. The key to a fully functional wireless platform that can enable solutions, such as retrofitting existing instrumentation for PAM, is to engineer and manage those networks from the top down.

Conclusion

A properly implemented wireless infrastructure combined with instrument retrofits allows for a quick and cost-effective implementation and provides plants and refineries the benefit of improved asset performance, availability and utilization. It's through the use of wireless communications that plants can now retrofit their intelligent wired industrial field devices to enable the trapped diagnostics, maintenance data, and even process measurements to have a low-cost path into an integrated asset performance management system. One of the lowest-cost and valuable solutions is to add a wireless adapters to the existing wired field devices allowing unused diagnostics to become part of a new or enhanced Plant Asset Management program.

It is widely understood in the industrial world that relying on degrading, failing, or poorly configured systems leads to higher operating and maintenance costs. Wireless retrofits coupled with the emerging wireless standards and a well executed and designed infrastructure, will usher in a new dawn in the operation of industrial plants and will bring new levels of productivity, uptime, and overall superior performance in the process industries.

Sources:

Hart Communication Website: http://www.hartcomm.org
ARC Advisory Group: http://arcweb.com

About The Author:

Paul Sereiko is the President and CEO of AirSprite Technologies Inc., a wireless sensing software and infrastructure company based in Marlborough, Massachusetts. He is also Co-Chair of the ISA SP100 Marketing Working Group and member of the HART Communication Foundation.

171f Base-Catalyzed Transesterification of Canola Oil in the Presence of Electric Fields

Zheyan Qiu, CPE, The University of Kansas, Learned Hall, 1530 W 15th Street, Lawrence, KS 66045 and Laurence R. Weatherley, Department of Chemical and Petroluem Engineering, The University of Kansas, Learned Hall, 1530 W 15th Street, Lawrence, KS 66045.

High voltage electric fields can enhance the mass transfer rates and phase separation in liquid-liquid systems. In our work the transesterification of canola oil with methanol was studied in an electrically intensified reactor using sodium hydroxide as catalyst. The reaction kinetics was investigated at different applied electrical fields. All the reactions were carried out at atmosphere pressure and temperature under 6:1 molar ratio of methanol to oil and 1%(w/w) by oil weight of NaOH. The results show that the conversion yields of methyl esters at the same reaction time were significantly enhanced as the strength of the applied electric field was increased. The increasing reaction rate mainly resulted from the enhanced interfacial area generated by electrospay and intensified mass transfer rate by interfacial turbulence in the presence of the electric field. The effect of the polarity of applied electric filed on the reaction rate was also discussed.

Virtual Simulation Environments using Online Control Software

Haruo TAKATSU*

* *IA Production Excellence Solutions Center, Yokogawa Electric Corporation, Tokyo, Japan, (Haruo.Takatsu@jp.yokogawa.com)*

Abstract: This paper describes a virtual simulation environment for peoples in the process automation industries. Objective is to re-use multiple online software available in the industry sites not only for operation and engineering but also for training and education purposes. At first the functional requirements are clarified. Secondly in order to integrate online software, process interface and human interface parts are focused on. Finally application examples are shown with concluding remarks.

1. INTRODUCTION

One of the major challenges facing Japanese process manufacturing industries is that employees including operators and engineers are decreasing due to the declining population and ageing Japanese society. On the other hand, the population of Asia is increasing and the growth of Asian factory is one of pressing issues to Japanese manufacturing industries. How to improve the operation performance of Japanese industries while minimizing the environmental impact and keeping the priority of their products over the global market, is a big concern of not only managing people but also Japanese people.

Generally speaking, an approach to improve the productivity and to increase the profit consists of three aspects: technology, systems and human resources.

At first the cutting-edge technology based products and services upgrades the competitiveness of the products and increase the profits as the result. But the effect of a new product will fade soon. So the sustainable submission of new products to the market is required.

Secondly systems approaches including marketing, sales, engineering, manufacturing, operation and maintenance are important to promote the business. A new business model will strengthen the sales of new products. To a specific issue of customers, the integration of tools, products and people can provide with the customer centric solutions efficiently and effectively.

The third is human resources. To innovare the human resource, the expansion of IT use, the provision of studies, the education programs for engineers are expected. Technical library based on the amount of track records through the development, implementation, engineering can be re-used at various sites all over the world.

As to the first aspect, we already presented some solutions shown in the reference documents. As to the third aspect, we will present near future, especially the activities of education program and knowledge library. So in this paper, one of the systems integration approaches as to the industrial process control systems is presented, especially focusing on the re-use of multiple online software.

The current process control systems constitutes the hierarchy of field layer, control layer, MES layer including APC and optimizer, MIS layer including planning & scheduling, based on ISA Standards. Figure 1 is an example of control system architecture, where each functional unit is connected with each other using the plant network such as Internet or Intranet. The progress of each function unit makes it possible to reuse it not only for plant operation but also training and education purpose without additional development. Using the process simulators instead of actual plants, other instrumentation systems, control systems and operation support software, et al. are available as are the same operation environments. We call this environment as 'virtual simulation and engineering environment' shown in Fig. 1, which is effective to evaluate the system performance and effectiveness before applying to the actual manufacturing plant in advance. Suzuki presented in INCOM2001 an example of the application of the dynamic simulator to the risk analysis of the industrial process. Takatsu presented in INCOM2006 the evaluation of online optimization using virtual simulation environment. The multi-variable model predictive optimizing control and online optimizer packages were integrated step by step to study and verify the effectiveness of the predictive control and online optimizer from the viewpoint of the total plant operation.

Our approach is to use the above virtual simulation environment not only for industrial process operation and engineering but also for the training and education purposes. In chapter 2, the functional requirements and specifications of each function unit are described at first. In chapter 3, a focus is put on the recent progress of interface technology, especially process interface and human-machine interface to integrate each function unit for industrial system. In chapter 4, some examples of education and training systems are introduced for validation purpose. Finally the concluding remarks are added on.

2. FUNCTIONAL REQUIREMENTS AND SPECIFICATIONS

To clarify the requirements of control systems, we should pay attention from the viewpoint of the total plant operation, the life cycle of the plant equipments, and user point of view. For example, Advanced Process Control (APC) should allow customers to operate its facilities with greater safety, cost effectiveness, reliability and compliance with environmental factors. Optimization is not a one-time event; it has to be a continuous effort to enhance operating performance in ever-changing conditions. Totally users should enjoy the benefits and improved profitability from the applied solutions. The next sessions show the requirements of each control function from users viewpoint.

2.1 Data Acquisition and Process Modeling

- **Easy model development** for process control: In a typical model-based Advanced Process Control (APC) project, 25% to 50% of the implementation time is spent in the process testing and model identification phases. An easy-to-use data acquisition and modeling make the intricate science of model identification available and saves the time of valuable project implementation.

- **Identification** technology is computationally fast. Its robust computation engine can process large dataset with unmeasured disturbance effects, and can identify models from both open- and closed-loop data.

- **Focus in efficiency**: Identification is designed to expedite the tasks performed during the course of a model identification analysis. Identification has a comprehensive reporting, project documentation, and "bookkeeping" functions.

- **Data Import**: Identification includes a versatile data import capability, designed to work with a variety of formats. Plant data can also be retrieved directly from data historians via the process variable retrieving tools.

- **Data Analysis** and Pre-processing: The powerful trending capabilities, data segmentation and bad data handling can be performed graphically. A wide variety of built-in calculations are also included.

- **Ease-of-use**: An identification workspace is a collection of process data, identification studies, and assembled models. It serves as a single-project archive and simplifies controller model maintenance. Identification tool is integrated with other process control software suite, sharing of objects between the various products, and also in its underlying modeling philosophy.

- **Incorporating Process Knowledge**: Users can perform constrained model identification using prior process knowledge. This can include setting the model parameters such as process gain, time constants, and dead time within bounds.

- **Grey Box Modeling**: The knowledge of cause-effect relationships can be incorporated during the process of model identification

- **Unmeasured Disturbance Modeling**: Rejecting unmeasured disturbances is one of the main purposes of closed loop control. Unmeasured disturbances include measurement noise, sensor drift, changing ambient identification, some controller parameters (for example, Gains), are incorrectly estimated. This could result in overly aggressive control action. Identification tool allows factoring in unmeasured disturbance effects during model identification.

- **Fit Statistics**: Statistical 95% confidence intervals for the estimated step responses and model parameters can be provided for use. Residual trends and goodness-of-fit tests, such as, residual auto- and cross-correlation analysis, can be used to check the quality of models.

- **Model Validation and Selection**: Several different models are built for use and their predictions can be compared against actual data. Models are easily constructed by mixing and matching various identification results. Final models are exported to controllers for online configuration.

2.2 Stabilize Your Plant and Maximize Its Profit

Multi-variable optimizing control strategies MPC, provides all the tools necessary to design, implement and maintain advanced multi-variable controllers. This is to improve the plant stability and maximize plant profitability for the hydrocarbon processing and chemical industries.

- **Squeeze and Shift**: MPC can be used to safely push a process unit towards its constraints area. MPC minimizes the distribution of key operating variables and shift the desired targets to the operating constraints, while maximizing the profit function. In case of Fluid Catalytic Cracking Unit, for example, the productivity is constrained by the Catalyst Circulation Rate. This constraint is not only holding back the throughput in the FCCU, but also restricts the overall refinery capacity.

MPC controller is commissioned to alleviate this constraint and increase unit throughput. The controller achieves the impressive improvement of around 1 ton/min in CCR. It raises the unit throughput and meet all the performance guarantees. It also exceeds the 95% average uptime required by the client.

Following control feature are also required: Easy to use control models/ robustness to unmeasured disturbance and unexpected model mismatch / nonlinear optimization / dynamic constraint handling.

- **Design**: The control dynamic models can be built either using the process acquisition data and/or a flow sheet and graphic builder. Offline configuration tools are also provided to design, test and build MPC controllers. The offline package generates the on-line controller implementation file. The offline package features a powerful simulation environment to test the controller behavior. Scenario based simulations can be used to test tuning, robustness to model errors and optimization performance.

- Online MPC part has functions for signal validation, controller initialization, mode shedding, standard operator and engineer interfaces and standard process interface using the de fact standard techniques.

2.3 Predict Product Quality for Better Control

When a critical product property is not measured online or only measured infrequently, quality estimation is used to predict the current value. An inferred measurement is typically calculated using selected basic process variable such as flow, temperature, pressure and so on. The quality estimation technology is used extensively in many refineries and chemical sites around the world. Typical applications in the hydrocarbon industry include:

-ASTM boiling point temperatures / Cold properties/ Flash point / Reid Vapor Pressure (RVP) / Road Octane Number (RON) / Impurity content in various streams / Melt index for polymerization processes

- **Why use quality estimator**: Quality estimator uses a unique online model calibration technique to make the predicted property value accurate and robust. It has a greatly expanded ability to perform intermediate calculations and include them as model inputs. Users can develop cascaded predictors and calculations within a single application, and can validate laboratory results prior to model calibration through an interactive data entry display.

- **Blending application**: Blending is one application area of quality estimation, as it needs any blending rules library or tools for tank quality estimation for use. It can also linearlize non linear models, such as blending rules, and pass updated gains to an MPC controller model.

- **Analyzer predictor**: Quality estimators give an early prediction of an analyzer signal to boost the performance of your multivariable predictive control application. It makes the control scheme robust against varying dead time, uncertain dynamics, and process non-linearity.

- **Virtual analyzer**: Quality estimator can be used as a replacement for an online analyzer, when it is not economical or when it is not technically feasible. Feedback from laboratory results is continuously used to keep the quality estimator internal prediction model accurate.

- **Design**: The estimator offline package is a Windows-based graphical package for the easy design and maintenance of inferred measurements. To create inferred measurements, users can select data analysis options ranging from multivariable regression to advanced modeling techniques such as Partial Least Squares (PLS) and neural networks. The estimator offline package also includes a configuration wizard for generating the online implementation file.

- **Simulation Tool**: It provides process control operators and engineers with the ability to operate the process effectively, to design a new application, to re-tune the existing applications or to provide with troubleshooting. Simulation makes it possible for human to compensate the weak points of control systems.

2.4 Real-time Optimizer

In order to utilize limited resources, it is important to take into account the environmental and safety issues such as the reduction of sulphur and lead, the restriction of CO2 generation, effective use of energy and so on. Real-time optimization calculates the best economic target values for process operation and control. In order to run the optimizer online and continuously, the following steps are dispensable. Refer to Fig. 2.

- **Data gathering and Collection**: Optimizer gathers the following type of data from the database such as hourly averaged data, raw data, constant data, constraint data, economic data, laboratory data and digital data. All data are collected as a background task and collected data are stored in historical data for the subsequent analysis and modeling purpose.

- **Unit Ready**: Before an optimization cycle begins, there must be confidence in the stability and range of the unit operation and the success of data collection. Stability and range of unit operation is measured in terms of Steady State Detect and Unit Configuration Check.

- **Steady State** is a measure of the steadiness of selected tag values. Optimizer determines on a statistical basis if the Unit process is steady at a particular moment. Steady state is achieved when, for all tags within a steady state set, the change in rolling average over 30 minute periods falls to within a specified tolerance.

- **Unit Configuration Check** tests whether the unit is operating normally, by checking a selection of tag values against normal operational range. It checks specific streams for abnormal operation conditions, e.g. the emergency steam flow to the regenerator is checked against a fixed threshold value.

- **Data Reconciliation**: Optimizer reads running hourly average process data and screen the raw data based on the low and high clamping values. If the data are good, it starts software for data reconciliation. Data reconciliation software uses model files to be able to reconcile the gathered data against the model rules, and writes reconciled data as output. Redundancy in measurement is used together with plant modeling to improve the accuracy of measured values and close heat and mass balances. Checking the convergence state of the result, it saves the reconciled data in the database.

- **Rating Case:** Using the data reconciliation output, the optimizer calculates missing process parameters like heat transfer coefficients for a heat exchanger and process bias parameters necessary for the predictive mode of the flow-sheet.

- **Base Case:** In the base case the model is represented in the predictive mode. This becomes a point of reference for measuring the changes resulting from the Optimize case and for measuring errors (bias) in the modeling. Constraint data measured is adjusted for bias. Process parameters from the rating case (e.g. heat transfer coefficients) are used and the actual control philosophy is translated into the model. The base case results are almost identical to the rating case results. Differences are mainly the result of numerical accuracy. Biases represent the difference between the model values and the measured values.

- **Economic objective function**: The optimizer objective is to maximize the profit within all relevant constraints, taking into account actual unit performance such as mechanical, operational and scheduling constraints. The economic objective function is a representation of the unit's marginal profits with respect to changes in feeds, products and utilities. The objective function is composed of three parts, total feed costs, total product values and total utility costs:

Objective Function = Product Streams Costs - Feed Streams Costs +/- Utility Streams Costs - other variable costs

Where: Each Stream Costs = Stream Flowrate (or Yields) * Steering Value

All feed, product and utility streams are degrees of freedom for the optimizer. Optimizer can change their flows and qualities and across the boundaries of the model, the value of a product stream is dependent on its composition and its eventual disposal. The product value if blended, will be corrected for its properties on the basis of primary quality with respect to the finished blend. Every time before running the optimization case, the economic objective function and underlying economic data should be confirmed.

- **Constraint**: Constraints can be process, operation and economic constraints such as valve opening range / pressure head of the pump upstream valve / normal operating range for safe and steady operation / DCS alarm limits / product quality constraint / equipment constraint / 'normal step' constraint; to limit the maximum moves suggested by the optimizer for safety reason.

Most of the constraints are dynamic in nature, such as the opening of the valve. Some are static in nature such as the products economics or technological unit limits.

By varying the unit operation parameters, input/output specifications or feed flow rates etc., we may hit the limits of the possibilities in the operation of the unit (alarm values), equipment constraints or quality constraints.

- **Optimization case:** It gathers real time controllers data required for the optimization run. It reviews the current process operation conditions and operates to maximize the profit using economic functions.

2.5 Improve Control Performance via Monitoring and Diagnosis

Generally speaking, APC implementation generates large economic benefits for users. But the achieved benefits are a direct function of the control system utilization and of its actual performance. Base layer controls, instruments and analyzers must also perform correctly to ensure high APC utilization and profit.

- What is required to monitor the controller performance

• Any type of commercially available multivariable controller can be monitored.

• Fast calculation and data handling allows a large number of loops to be analyzed in a short period of time.

• Diagnostic measures and statistics, such as closed loop response speed, can be compared with other benchmarks.

• The applied technology and user interface are easy to use and understand.

• Monitoring reports can be shared using standard MS Windows applications.

- How to compute Performance and Benefits Statistics

PONC (price of non-conformance) is the cost of lost opportunity associated with process or controller downtime. PONC is the difference between the maximum daily benefit and the realized daily benefit.

Uptime ratio: The amount of time that a particular controlled variable (CV) is "on control" during the day.

In-service ratio: The amount of time during the day that a CV was operational or available for control.

In-compliance ratio: The amount of time during the day that a CV was on control and within a pre-specified tolerance of its set point or set range.

- What is output ? A summary of control performance for the entire facility, with uptime, compliance and in service statistics presented for each process unit.

Bad actors: Two lists of the worst performing CVs in the facility ranked by consecutive days of not meeting objectives. The first list applies to CVs that are in-service, and the second list applies to CVs that are out-of-service. A list of the controlled variables in a unit that are not meeting user defined control performance objectives.

On-demand reports: Customizable summary reports consisting of all control performance and benefits statistics over a user-defined time period.

3. INTEGRATION TECHNOOLOGIES

In order to integrate a variety functions to construct total systems, the communication interface is a key technology. It is required to be open, flexible and promising to application in all aspects. These days, OLE for Process Control (OPC), Web Interface have recognized as de-facto standards. Simulation technology has become commonly available in almost all fields of academy and industry. How to describe the simulation models and how to interface between simulator and other application should be standardized.

3.1 Process Interface

In the process and manufacturing factories various equipments, devices, instruments are provided by each vendor. OPC is a real-time data exchange techniques based on the standards of OPC Foundation, and has become popular and applied lots of DCS, PLC and PC beyond the difference of manufacturing vendors as common process input and output standards. Refer to Fig. 3.

- OPC DA interface is applicable to exchange the current real-time data among different applications, such as process input/output, data calculation and data exchange. In order to connect with different vendors products, compliance to the OPC standard is indispensable, such as OPC quality handling, error processing, performance, cache/device read mode, et al.

- OPC HDA and AE interfaces are available as common access for HMI trend data, alarm/event messages and backup.

- To exchange data among multiple OPC servers, OPC based data exchanging is useful for industrial applications. Error handling in abnormal situation, configuration easiness by graphical builder and debugging function are also indispensable. Refer to Fig.4.

- Near future, the text based OPC interface such as OPC-UA will become popular to improve the connectivity and to expand the openness of the information. The more progress of semiconductor and communication are expected.

3.2 Human Interface

Web-based technology can be applied on every aspect of manufacturing industries, such as planning, operation, control, management and maintenance. Because users expect 'thin client' such as almost no client software installation, no copy of application files and not setup of implementation. Recent progress of Industrial Ethernet technology accelerates this trend more and more. On the other hand, to secure the target system from unexpected access, security setting becomes important more and more. Web based HMI is also requested in the manufacturing industrial system. But before implementation, we should clarify what kind of functions is requested and who will use these functions. Below show what functions to be required, who to require and when it is required for Web based HMI .

<Functionality> Operation anywhere: monitoring and operation anywhere such as from other domain, from outside firewall, from outside office, from Internet or Intranet, et al.

Integration: integrating the information of multiple servers such as acquisition data, APC data, optimization data et al.

Security: protects the data from unauthorized level or miss operation to provide secure and safety environments.

<Role> Expected staff of the system is as follows: Field engineer / Operator / DCS engineer / Lab engineer / Control engineer / maintenance engineer / supervisor / manager / director/

<Time chart from the viewpoint of plant life cycle> Expected plan life cycle is as follows: Design / plant test / modeling / implementation / operation / maintenance. Table 1 shows an example of the time chart of each operation.

Otherwise, not only Operator windows but also definition files between offline and online, debugging files are standardized using XML format.

3.3 Simulator Interface

Online use of simulator will become available to support prediction, optimization, scheduling of plant operation. So the following specifications are recommended.

- Simulator should have the same interface as the process interface to simply change the connection between simulator and actual process without changing other applications. For example OPC Interface is used to access data from/to simulators.

- The simulator can be started and stopped from the remote PC. After starting the task, the execution processing is done in the fixed cycle by the scheduler. The task is started and stopped from the simulation package. In addition, the execution processing is synchronized with the simulation package. In this case, by performing the execution processing request from the simulation package, the task can be synchronized with the simulation package. Refer to Fig. 5.

4. APPLICATIION EXAMPLES

Figure 6 is an example of Web based HMI. Multiple controller information from multiple servers is displayed in one window to realize 'operation anywhere'. Figure 7 is an example of monitoring and diagnosis report to manage the control performance of all the sites for 'information sharing'.

5. CONCLUSING REMARKS

Virtual simulation environments concept is proposed not only for process operation and engineering but also for the training and education purposes. By standardizing the interface parts, lots of existing online software are available beyond the difference of company and academy globally.

REFERENCES

Takatsu, Okada, Nozaki (2006). *Evaluation of Online Optimization Using Dynamic Simulation Environment*, IFAC Symposium on Information Control Problems in Manufacturing, INCOM2006

Takatsu, Okada, Nozaki (2004). *Application of Simulation Technology to Plant Optimization and Control*, SaA03 2004/CCA/ISIC/CACSD

Takatsu (2007), *Advanced Process Control and Optimization Solutions*, Yokogawa Technical Report No.43

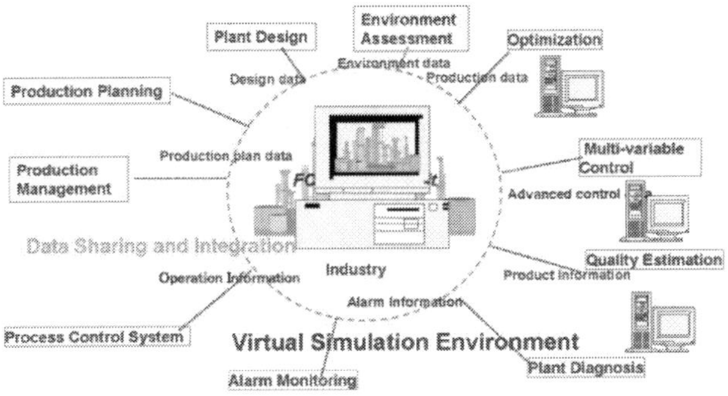

Fig. 1 Virtual Simulation Environment

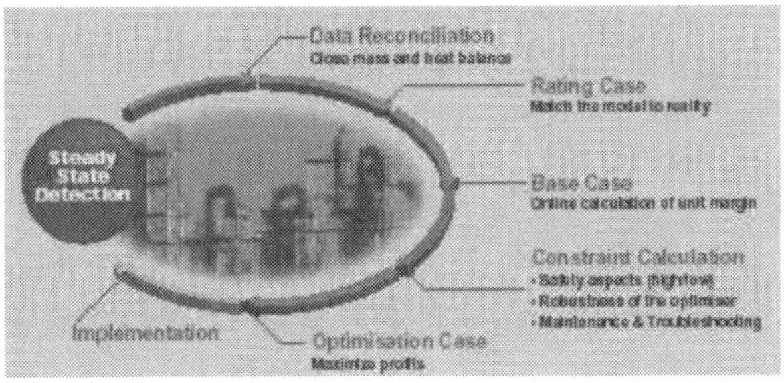

Fig. 2 Optimization cycle

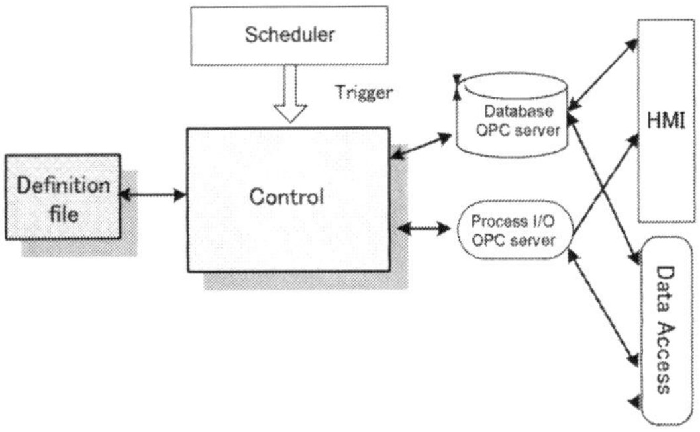

Fig. 3 OPC based process interface

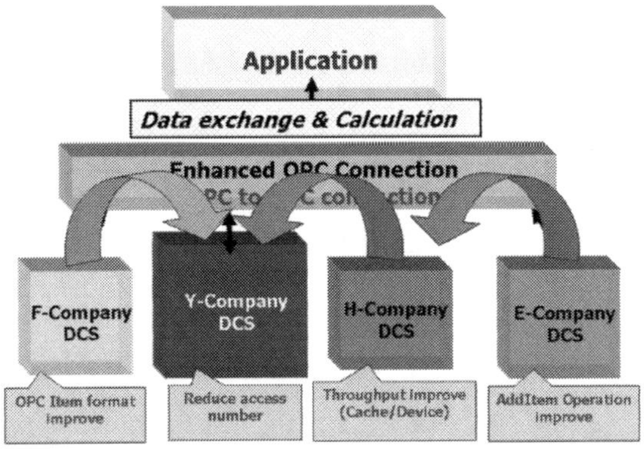

Fig. 4 OPC based data exchange

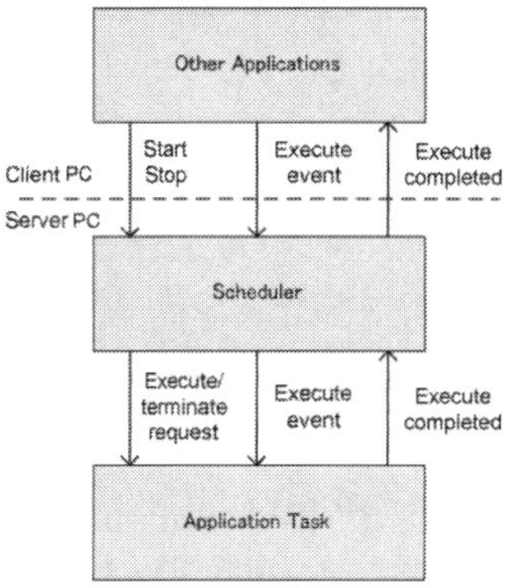

Fig.5 Simulation Interface

Table 1 Example of role map in each phase

Phase	Engineer	Operator	Manager
Planning	Remote access		Integration
Design	Remote access		
Implementation	Monitoring File exchange	Training, Education	
Operation	Monitoring	Monitoring Operation	Performance Management
Maintenance	Monitoring Diagnosis	Monitoring	
Improvement	Operation, monitoring		Performance Management
Evaluation	Remote access Monitoring		Monitoring Integration

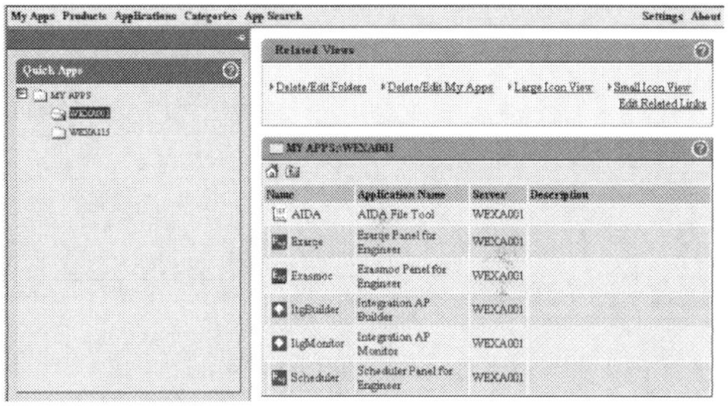

Fig.6 Example of Web based HMI

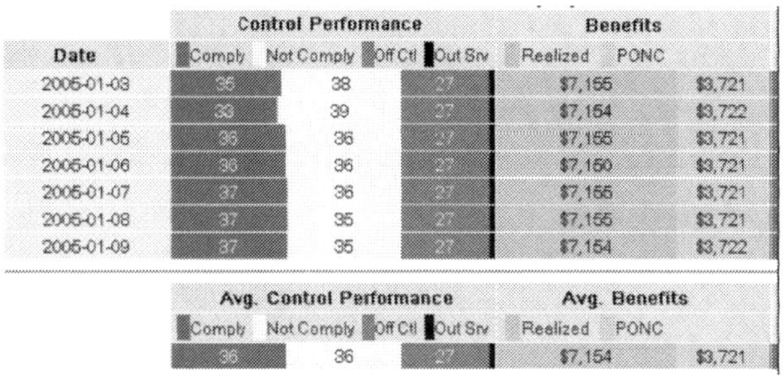

Fig.7 Example of Monitoring and Diagnosis function

73a Plant Troubleshooting: Engineers as Detectives

Shawna S. Berg, Process Engineering and Evaluations, Shell Global Solutions (US) Inc., Westhollow Technology Center, 3333 Highway 6 South, Houston, TX 77082

It's 2am and the plant calls to tell you that the unit is not running well and they have no idea what is wrong. All eyes are on you to figure out what happened. Where do you start? How do you find the information you need? As engineering majors, not only were we taught how to design heat exchangers and run a material balance, but we were also shown how to solve problems. This paper outlines the mental process used to troubleshoot plant issues including the information that needs to be gathered to solve the problem and the role an engineer plays as a liaison for the team.

233d Fundamental Liquid-Liquid Equilibrium Model for Asymmetric Component Systems

David Denton, Plastics R&D, The Dow Chemical Company, 2301 N. Brazosport Blvd., Building B-4001, Freeport, TX 77541, Prasanna Jog, MI Analytical Sciences, The Dow Chemical Company, 1897 Building / F37, Midland, MI 48674, and Li Max, E&PS Reaction Engineering, The Dow Chemical Company, 2301 N. Brazosport Blvd., Building B-1603, Freeport, TX 77541.

In multi-component miscible systems where homogeneous single phase solution behavior is critical, it is necessary to model and predict regions of instability through the process. Changes in the temperature, pressure, and composition of multi-component systems may cause separation into two distinct liquid phases. This is detrimental in some unit operations such as mixing, reaction, and heat transfer, but is advantageously used in some unit operations such as separations. Knowing the boundaries for thermodynamic phase equilibrium behavior becomes important as processes are optimized. We report an application of the improved technique for modeling, using Perturbed Chain-Statistical Associating Fluid Theory, which better incorporates the asymmetrical multi-component thermodynamic phase solution behavior. As process conditions change through these unit operations, the system temperatures and pressures can be controlled as needed for each composition to avoid the two phase separation of the Liquid-Liquid Equilibrium. In one demonstration, this application was used to optimize the performance of a heat exchanger by predicting and controlling the point of Liquid-Liquid Equilibrium for the process through the exchanger tubes.

12D00 Plant Troubleshooting and Problem Solving: The Chemist and Engineer Working Together Tolling Success: Your Chemists and Our Engineers Working Together

William G. Baker – willb@kmcoinc.com, Business Development – Custom Processing, KMCO LP, 333 N. Sam Houston Pkwy., E., Houston TX 77060

Chemists drive the development of new molecules, but the transition from bench to plant will make or break a project. When a Toll Processor is used to prove out the process or to get to market sooner, the production scale-up can be complicated by corporate boundaries, limited communication, tolling schedules, and tolling equipment built with only general production capabilities. Communications determines the success or failure of the final project. This session will highlight chemists and engineers working together in a tolling environment. Case histories, examples, and best practices will be presented.

The customer's Chemists and the manufacturer's Engineers work together typically under two scenarios; project startup and problem resolution. Often, there is an inherent conflict between the theoretical nature of the chemistries involved and the actual circumstances surrounding the project. In theory, under a certain set of conditions a process will yield a particular result. In actuality, conditions may be compromised by equipment and resources, impurities in the materials, or variances in the execution of the process. Depending on the project, the difference can be inconsequential, or it may lead to profound challenges to project success. When challenges arise, it is important that all the involved parties remain focused on *what* is happening and *what* can be done rather than *why* it happened and *why* something was or wasn't done.

The Point Where the Chemist and the Engineer Must Work Together: The Genesis of a Project

A market need is identified and solution is hypothesized. You know the chemistry, you can source the raw materials and the project will be a success; at least it is on paper. Or, the economics dictate a regional manufacturing approach is best for the growth of a particular project and your plant resources are best utilized elsewhere. Or, you require an alternate manufacturing source to satisfy market concerns with supply. There are numerous circumstances that drive chemical companies to Tollers to manufacture their products, and the level of success of every project is always representative of how satisfied all parties are with the final results. The deal always looks good on paper (at least it should) as the business and technical representatives of both parties have laid out a blueprint for project success. Whether or not that comes to fruition will be determined by how well the chemists and the engineers work together going forward.

At this point, if you think the negotiation is over, think again. Anytime two people exchange information, they are negotiating. When you negotiate, you share information to change the relationship. When you share information in a positive way, you get a positive result. Trust is vital to a negotiation. If people don't trust you, they are not going to share information with you. The confidentiality of a project and its technical merits are paramount and understood by all those involved. Even so, each parties representatives must earn one another's confidence in a short period of time to maximize the project's chances of success.

Reaction synthesis is more than following a chemical recipe, there are always particular aspects of the process equipment, or analytical laboratory resources, or logistics functionality of the tolling facility that must be accommodated along the way to processing success. This is because very rarely is a tolling facility designed for a specific process or procedure. If the facility is new, the assets are designed to accommodate the initial chemistries that support the business, but over time new opportunities expand the toller's capabilities. Equipment is usually designed to accommodate a generic range of chemistries and volumes. The analytical laboratory is much the same with a range of in-house capabilities and known outside resources to supplement the labs ability to analyze materials. The inherent logistics of a tolling facility can affect the project by how which the materials are handled. I am not advocating that the chemist or the engineers be in control of all these aspects that may or may not affect the process, but they must recognize the challenge areas are critical to a smooth running project.

Maintain Focus on the Project, Not the Personality

It is important that when the chemist and the engineer communicate that they maintain their focus on the project and its success rather than the people and personalities involved. Know why it is that you are there in the first place, and have an understanding what motivates the party across the table. For the chemist representing the customer, he/she is there to make sure the product can be manufactured on specification in a reliable manner that will satisfy the current scope of the project. For the engineer representing the manufacturer, he/she is focused on developing a process that will minimize the project expense while satisfying the customer's needs. For example, when a project starts up and the customer's chemists are on site overseeing the first production run and concerns are expressed about the equipment and the facility in a negative way, the manufacturer's engineers assigned to this project are going to limit the information they will expose to such criticism. The level of project success is immediately compromised because the key technical personalities have clashed. Though the toll manufacturer may have promoted itself as an extension of the customer's own facility, it is not, and must be respected as such. Rather than express concerns about the equipment and facility in a negative way, understand the how the manufacturer plans on utilizing their equipment to manufacture the product. Help the manufacture understand what your concerns are that are specific to the project at hand and understand their explanation. You may have to trust in their experience running their equipment, but when your concerns are expressed with confidence that the manufacturer will have the answer that will satisfy your concern, you will get a deeper understanding of the equipment and facility that will allow you to make better decisions in the future. If the communication between the technical representatives on the ground is side tracked by accusatory comments, the material still has a chance to be manufactured to the proper specification, but it won't be because the chemist and the engineer worked to make the process run any better. The mutual understanding of these motivating factors through the proving process ensures that both parties will be able to stake a claim in the project's success.

For example, in a situation where something has gone awry in manufacturing that leads to hundreds of thousands of pounds of off-spec material, how well the customer's chemist and the manufacturer's engineer work together will determine the expediency in which the matter is resolved. Don't begin the reconciliation by pointing out who is to blame and why it happened unless there was never any plan for reconciliation. This will be a waste of time where the relevant personnel will take a defensive posture and defend themselves from further criticism.

Even if the fault is clear, establishing who the party at fault is will get nothing done to remedy the problem and one of the problem solving resources will likely not be as helpful as they could have been. Point out what went wrong and establish that you are not interested in who is to blame as much as you are interested in collaborating to work out the issues in a manner that will benefit both parties. It is in everyone's best interest for the issue to be resolved as soon as possible and have measures put in place in an effort to prevent the problem from happening in the future. Maintain focus on solving the problem and not assigning the blame. Determine what happened and what the next step will be.

The customer's chemist needs to listen to the manufacturer's engineer's explanation of the situation and focus on identifying the best chemical processes that will fix the problem (if indeed it can be fixed). The manufacturer's engineer needs to listen to the customer's chemist and focus on the best execution of the process that will fix the problem. Understand the issues outside the chemistry and equipment that may also be pressuring the process such as time and cost. When the dialogue focuses on issues and solutions independent of the people, relevant information will be forthcoming and better decisions will be made sooner rather than later.

Ask Questions, Listen and Understand What is Meant

Ask questions then listen and understand what is said and make sure you understand what is meant. Many times we come from different professional and cultural backgrounds and we may interpret what we are told much differently than what was meant. Just as dinner time is at 6pm in some places and 12 noon in others, everyday in one facility might mean Monday through Sunday, and in another facility it may mean Monday through Friday.

Even if you know the answer, always ask the question. Often, when we answer questions, we answer the wrong questions. Always make sure you understand the question. Ask "what" questions, not "why" questions. Never argue, but always question for understanding. This is when you want to ask "what" questions. If you really want to change the other person's mind you have to educate them first. Ask the other person what they base their figures, analysis, etc. on. Don't tell them the information is ridiculous. Educate them through asking questions.

In this ISO certified world, we trust the other parties are certified and competent in their duties. You will understand people better when you witness how they handle the details. When you are working with someone for the first time on a project, work through every detail of the process together regardless of how standard or mundane the process is. This will allow both parties to understand the project details that will develop mutual confidence in the understanding of what will make this project work best for both parties.

When proving out processes or working through problems, a mutual understanding of all issues is paramount to project success. Years of continued education and field experience can make you very smart in your field, but when you are exchanging information during project start-up or troubleshooting a process, the dumbest things you can do is act smart and not hear each technical point communicated back to you from the other side. Sometimes, you will find that the initial understanding was not mutual and further discussion prevented wasted time or worse yet, ruined products.

For example, a customer comes to a manufacturer for the first time with a project to make Glycol Ethers by ethoxylating an alcohol with Ethylene Oxide using Sodium Hydroxide as a catalyst. This is simple in nature, and a facility that specializes in ethoxylation should be able to handle this simple project, right? Well if the manufacturer was already producing the exact same product, to the exact same specification using the exact same raw materials, then yes, you should be able to take them at their word and enjoy project success. But this is not likely the case.

This particular product specification is tight and the alcohol material ranges in purity and water content. The customer manufacturers this same product in their own plant where they are planning on reallocating the assets. They need this product to be toll manufactured. The customer offers to have the manufacturer's engineers take a trip to their plant and learn how the customer manufactures and handles this material.

The customer not taking the time up front to understand the manufacturer's expertise and capabilities first hand could lead to months of lost time and hundreds of thousands of pounds of off-spec material. Though first hand knowledge of the customer's process would give the manufacturer insight into the subtleties of the process and the steps the customer emphasizes in their own process, there is a way that the same type of information can be understood by the manufacturer.

In this case, the manufacturer assures the customer that they know how to make these molecules that they make these all the time and the project will be a "piece of cake", and decline the offer to see the customer's process first hand as they feel it is not necessary.

The customer instead should insist on having their chemist take part in the setup and the first trial run to ensure the project gets started with the best chance of success. Having experienced eyes evaluate the material throughout the process will minimize processing uncertainty, and in the event the initial processing runs produce non-conforming material, the customer and the manufacturer share the responsibility and can act effectively to get the process on track.

Though the manufacturer ultimately determines what happens in their plant, the raw materials usually belong to the customer and having representatives overseeing their investment should be mutually understood. The customer's chemist should ask questions at every step in the process to verify his understanding of not only each step in the process but also the manufacturer's interpretation of that same step. This line of questioning should generate discussions where the chemist and the engineer share details that will ultimately support the success of the project which is mutually beneficial.

Principals of Constructive Communication: Separate the people from the problem

If you see the person as the problem, there will always be a problem. Your problem is not with the person, it is with their thinking. Focus on their interests, not their positions. A person's interests are what cause them to develop their positions. When working through an issue with the personalities involved in the process, find out what it is that they want. That is where you will usually find common ground.

Let's say that during the startup of a new project, the process is running slower than expected because the process temperatures are not being achieved in a timely manner and the manufacturer's engineer takes the position that there is no other option and the process will have to be run with slower than expected heat-ups. Rather than threatening the manufacturer with the possibility of non-conforming product due to the slow heat-up and focusing on why the engineer would settle on such a position, explain the chemical reasons that make the heat-up critical to the process and share what you expect the outcome to be. Rather than hope that the material comes out on-spec, ask questions to get an understanding of what the barriers are that are preventing the heat-up. Without making judgments regarding the state of the equipment or execution of the project, make it clear that your interest pertains to the success of the project and begin discussions that lay out the different scenarios based on the plausible outcomes from processing under the compromised conditions. Make sure you have a mutual understanding of exactly what is happening in the process and let the business representatives handle the reconciliation if necessary. This will leave you with an unimpaired technical working relationship that will benefit possible future projects.

Principals of Constructive Technical Communication: Generate a Variety of Possibilities when Working Through Issues

One thing that helps you establish credibility at the outset of a negotiation is to talk about all the options the other party has, even the ones you do not want them to take. Bring up the last thing you want them to do first. This will get it on the table so they can't threaten you with it. If you're not talking about it, they are thinking about it and thus are not listening to you. This tactic will also help establish you as trustworthy. After you discuss the options, start to talk about the consequences. If you bring up the worse options from your point of view, you can put your own spin on it. This can bring up aspects of that option they had not considered.

When confronted with technical issues that threaten the viability of a project, generate a wide variety of possibilities before deciding what to do solve the problem. Having a lot on the line in a manufacturing project will inhibit creativity. So does searching for the one best decision to work through an issue. Once you decide on a resolution path, continue to understand how the new plan will affect both parties and explore the details of the process because the other party in the project might be able to help you out in ways that wouldn't occur to you.

For example, the processing of a crude chemical stream is jeopardized by large amounts of solids periodically appearing in the feed which are finding their way into the processed product. The source of the solids ending up in the product is unknown to both parties. Rather than accuse the manufacturer of poor material handling or the customer of supplying bad material, the chemist assigned to the project by the customer determines that the feed can be treated with an inhibitor prior to processing that will keep the solids in solution throughout the processing. Now, the manufacturer's engineer suggests that they can enhance the effectiveness of the inhibitor by moving the feed to a tank with a high suction line allowing the majority of the heavy solid forming compounds remain in the tank. The engineer also suggests that they inject the inhibitor in process through a by-pass feed system that the manufacturer uses for a similar project. The collaborative solution works better for the customer who benefits from better utilization of the additive resulting in consistent product as well as the manufacturer who does not have to be deal with pouring the additive into the tank and taking the time to agitate it to assure thorough mixing. The added benefit to both parties is the by-pass injection system allowed the customer to source the inhibitor more economically and moving the feed tank allowed the manufacture to process larger, more efficient campaigns. Because the chemist and the engineer remained focused on the task at hand, the best solution was identified and implemented.

Principals of Constructive Technical Communication: Preparation for the Communication

Give a lot of thought to how you prepare the agenda. Determine the issues you need to negotiate. If you don't know what is important to the other party, ask them to prepare the agenda. If you don't want them to prepare the agenda, ask them to write you a memo listing all the things they want to make sure you talk about. Then prepare your agenda independent of that information. Once you get the other party's memo, look for things you put on your agenda and things you had left off your agenda. The way you make things non-negotiable is by keeping them off the agenda.

More often than not, you don't want to surprise the other party. So, send them the agenda beforehand. You are not looking just for a solution, but for a solution that will last. A solution that won't stand up to close examination will not last, but will be the source of further conflict and negotiation.

When you can control time, information and climate, you can develop options. But it is very difficult to drive the development of options unless you have done your homework about the other party and developed a plan.

First, examine the needs and deadlines of the other party as well as your own. When a technical issue arises in a process, you know what your needs are regarding resolution, but if your goal is the best solution in the shortest amount of time, identify and understand the needs of the other party as well. Each others needs are usually ties to time deadlines, and sharing and understanding each other's time frames will enhance the available options for resolution. Only when the solution satisfies both parties do you have the best solution.

Second, list your assumptions and validate each one. Depending on the situation, your assumptions may include equipment types, analytical methods, and other technical details germane to the project. Sometimes your assumptions surround the sources of power and influence that are driving both parties. Be prepared to validate assumptions that may appear to be clear such as product specifications. You may find in some cases even these terms are flexible and may be the means that lead to production success.

Third, make a decision tree. This is best done in a brainstorming session with your group. Ask "If I say this, what will/can they say?" and be prepared to respond appropriately. Being prepared to respond to the other parties questions and responses will give you the best chance to drive the discussions in the direction that you choose, with the other party following along willingly.

Principals of Constructive Technical Communication: Impasse

When you have reached an impasse in a discussion of the technical execution of a new project or determining the solution to an existing problem, take a step back and give yourself a different perspective to the situation. First, check your approach. You may be forcing too much, or you may just not have asked for an agreement. Next, check the benefits. The more you know about something, the more you are likely to talk about the properties and processes you are familiar with instead of how your solution will benefit the other party. Tell them what the benefit is to them. Finally, check the people. You just may not have the right people on your team with the understanding or authority to move the project forward.

Conclusion

Begin every technical communication with the belief that there is a solution to every technical problem that will benefit both parties. Maintain focus on the project and not the person. Ask questions, listen and understand what you are being told. Separate the people involved from the problem, generate a wide variety of possibilities surrounding an issue, and prepare thoroughly for communications. This will ensure that you will control the positive exchange of information that will lead to a positive outcome.

171a Performance of An Intensified Downdraft Gasifier at 50kWe and 1 MWe Scales

G. Akay[1], B. Calkan[1], O. F. Calkan[2], F. J. Herrera[1], C. A. Jordan[1], A. H. Mohamed[1], and M. Tham[1].
(1) School of Chemical Engineering and Advanced Materials, Newcastle University, Merz Court, Newcastle upon Tyne, NE1 7RU, United Kingdom, (2) ITI Energy, Innovation Technology Centre, Brunel Way, Rotherham, S60 5WG, United Kingdom

An intensified downdraft gasifier system capable of generating 50kWe (with turn-down ratio of 5:1) has been developed recently and subsequently scaled up to generate 1MWe and 1.5MWe by ITI Energy. Extensive experiments were carried out both at 50kWe and 1 MWe levels in order to assess syngas composition and calorific value, exhaust and process water emissions and ash composition while monitoring line pressures and temperature distribution in the gasifier reactor. After syngas cleaning, electricity was generated using an internal combustion engine. Biomass fuels used were: municipal solid waste, wood chips and bone meal at 1 MWe plant while at the 50kWe capacity, additional fuels included fuel cane bagasse and refinery sludge. Performance of the gasifiers was compared as regards syngas composition, temperature profile, tar and particulate levels. The gasifier reactor was modelled in order to predict the temperature distribution and syngas composition while developing a control strategy for the gasifier.

Web Page: www.iti-energy.com and www.Newcastle.ac.uk/pim

206a Overview of Steam Trap Types and Their Operation

Norman A. White, TLV CORPORATION, 13901 South Lakes Drive, Charlotte, NC 28173

Steam traps are the quietly functioning devices underneath the pipe racks of our manufacturing facilities, giving the occasional breath of condensate/steam to the atmosphere. However they are not all created equally or function the same way. This presentation will give an overview of what types are out there and their pros and cons.

206b Process Improvements Via Steam Traps Design and Configuration

Norman A. White, TLV CORPORATION, 13901 South Lakes Drive, Charlotte, NC 28173

The most obvious benefit of a healthy steam trap population is the money saved on condensate returned. Often overlooked are the process improvements gained from proper steam trap selection and design especially around exchangers. This paper will discuss various arrangements and their benefits to the process.

206c Practically Managing Your Steam Traps Including the Common Pitfalls and Mistakes

Jason V. VanRoeyen, Geismar Manufacturing Facility, Shell Chemical Company, River Road, Geismar, LA 70734

At times it appears the only companies that value how important healthy steam trap populations are the ones that sell steam traps. This is an unfortunate side effect of minimal resources to monitor steam traps. This is due to cost cutting and lack of clear ownership for steam traps once inside the operating unit battery limits. This topic will explore the realistic view, pitfalls, and operation of steam traps at a manufacturing location.

206d Strategies in Optimizing Condensate Return

Debbie Bloom, Nalco Company, Naperville, IL 60540

Optimizing condensate return for reuse as boiler feedwater is often a viable means of reducing fuel costs and improving boiler system efficiency. Condensate that is contaminated with corrosion products or process chemicals, however, is ill fit for reuse; and steam or condensate that leaks from piping, valves, traps and connections cannot be recovered. A document published by the US Department of Energy estimated that for all pulp and paper, chemical manufacturing, and petroleum refining industry plants1, approximately:

*66% could realize typical fuel savings of 3 to 7% with an effective steam trap management program

*6.5% could realize typical fuel savings of 2.9% by minimizing vented steam

*24% could realize typical fuel savings of 2% by optimizing condensate return

*16% could realize typical fuel savings of 1.4% by repairing steam leaks

*7.8% could realize typical fuel savings of 0.9% by isolating steam from unused lines

As more condensate is returned, less makeup is required, saving on water and makeup water treatment costs. The high purity of the condensate allows for greater boiler cycles of concentration, thus reducing water and energy losses to blowdown. The high heat content (typically in excess of 180°F) can provide substantial energy savings. Additional savings will also be noted in reduced water treatment chemicals, water and sewer costs. Effective chemical treatment, in conjunction with mechanical system improvements, condensate polishers, and automatic dump systems can assure that condensate can be safely returned and valuable energy recovered.

1U.S. Department of Energy, "Steam System Opportunity Assessment for the Pulp and Paper, Chemical Manufacturing, and Petroleum Refining Industries," DOE/GO-102002-1639, September 2002.

73f Lean Six Sigma Approach in Chemical Process Troubleshooting

Zhen Lai and Chieh-Min Cheng. Xerox Corporation, 800 Phillips Rd, Bldg 143-02S, Webster, NY 14580

Lean Six Sigma is a proven process improvement methodology currently widely being employed across nearly every type of business and industry including numerous chemical companies. This paper describes the application of Lean Six Sigma methodology for manufacturing problem solving in a chemical plant. Examples and case studies for real chemical manufacture process trouble shooting will be described. A systematic problem solving process, DMAIC process (Define, Measure, Analyze, Improve and Control), was demonstrated from the point of view of the practicing chemical engineer.

25f Studying Water Quality by Corrosion & Scaling Index with Changing the Method of Microorganisms Control

Ehsan Bakhshi, R&D Center of National Petrochemical Company, Tehran, Iran and Fatemeh Abniki, Process Eng.Dept. of National Petrochemical Company,Ghadir Group, Tehran, Iran.

The use of biocides in water cooling towers, specially in open recirculating systems to prevent problems arising from the microorganisms is an important subject. With the aim of removing non oxidizing biocides in these towers, a research project was started . During the 16 months period, sodium hypochlorite was injected to the system. Production of sodium hypochlorite in the complex made its selection as a substitute for biocides, logical and cost effective.

According to a regular program, three times a day, sodium hypochlorite was injected into the basins of cooling towers so that the extent of free residual chlorine was between 0.3 - 0.6 ppm. The results of microbial and chemical tests during the control period indicate that TBC reduced to less than 30 cell /ml (allowable level). The existence of SRB was mostly negative.

Corrosion Results & Ryznar Stability Index

The test conducted by using corrosion coupons indicated that corrosion rate was less than 1MPY most times. The highest corrosion rate, 3MPY , belonged to LD plant . Apart from coupon test , Ryznar Stability Index ,RSI, Which determines corrosive or scaling tendency of water was calculated in 13 months.

Using the available sources and the conducted studies, the optimal range of RSI was found to be between 5.5 to 6.5, in that case water has neither corrosive nor scaling tendency. Results show that seven out of the eleven plants, (CA , UT , AR , BD/SR , VC , OL , and NF) the RSI monthly mean index had constantly remained at the optimal level of 5.5-6.5 during the control period . It is important to note that no month's average was within the corrosive or scaling limit.

In PVC plant 92 percent (12 months) of monthly mean indices were inside the appropriate limit.

LD plant with a monthly average RSI of 70 percent within the optimal range, had the lowest standing among plants. PP and HD plants, 77 percent of the time were in that range.

Generally , considering that every plant achieved more than 70 percent of the suitable level on Ryznar Stability index , also, with a view of coupon test results , it seems that the results of corrosion and scaling control in water cooling towers had been successful, with changing the method of microorganisms control.

Web Page: www.nipc.net

Effects of Supercritical and Subcritical Pressures of Water on Coke Formation During Jet-A Reformation to Hydrogen

Michael S. Stever, Jason W. Picou, Jonathan E. Wenzel Satya Putta, H. Bryan Lanterman[1] and Sunggyu Lee, Missouri University of Science and Technology, Rolla, MO

Abstract

Jet-A, a civilian aviation fuel, was reformed to hydrogen using supercritical water as a novel reacting medium that acts both as a solvent and as a reactant. Supercritical water has also been demonstrated as an effective reformation medium for a variety of hydrocarbons. This experimental system is scalable with reformation of hydrocarbon feedstocks demonstrated in a 400 mL Hanyes® Alloy 230 tubular reactor and a 1-L Inconel® 625 Grade 1 alloy tubular reactor. Due to compact reactor volumes this process can be fabricated into portable power units. The experiments discussed in this work were conducted in a 1-L Inconel 625 Grade 1 alloy reactor with temperatures ranging from approximately 923 to 983 K. Pressures for the subcritical experimental runs were investigated from 17.23 to 21.99 MPa while the supercritical studies were examined in the range of 23.49 to 32.54 MPa. Coke residues were analyzed for percent composition of carbon and hydrogen. Mass of fuel fed was compared against total coke mass and compared with normalized carbon and hydrogen composition analyses from the various cokes to examine the effects on the composition of the coke products. This paper provides a quantitative analysis of hydrogen and carbon content in coke formed during subcritical and supercritical water Jet-A reformation. The result provides evidence for the significance of the supercriticality of the input process condition on the supercritical water reformation process.

Introduction

On-site generation of hydrogen from a variety of feedstocks, both renewable and nonrenewable, is an essential step in moving towards a hydrogen-based economy. A novel process presently under development to produce hydrogen is the supercritical water reformation of petroleum-based fuels in which supercritical water acts synergistically both as a solvent and as a reactant.[1] This process is non-catalytic thereby eliminating the problems inherent with catalytic reformation including catalyst fouling due to coking in the reactor, catalyst deactivation due to sulfur and organonitrogen poisoning, and catalyst sintering caused by autothermal process heat.

Carbon and hydrogen are present in three different phases in the reactor: (1) as diatomic hydrogen and gaseous species containing carbon and hydrogen, (2) liquid species containing carbon and hydrogen, and (3) coke residues whose molecular structures are largely deficient in hydrogen atoms. The primary interest for this work is the carbon and hydrogen present in solid coke residues, which are resulted from

[1] Currently with DRS technical Services Inc., Alexandria, Va.

supercritical and subcritical water reformation experiments. As part of an effort to minimize coke formation in supercritical water reformation of jet fuel, a study was performed to determine the effects of reactor pressure upon coke formation and coke polyaromicity.

Background

Jet-A, a civilian aviation fuel, is a petroleum distillate fuel of a complex composition of various chemical species. Straight chain hydrocarbons ranging from nine to fourteen carbons comprise approximately half of the weight percent of the fuel. Aromatic and polycyclic aromatic hydrocarbons (PAHs) of varying orders of polyaromaticity make up the rest of the fuel. [2,3] Using ASTM Method D2887 for boiling range distribution analysis, the largest probability density for Jet-A was found at a carbon number equal to twelve as seen in Figure 1.[4] Based on this result, a single representative molecular species for Jet-A fuel may be assumed to be n-dodecane.[5] Sulfur is also present in the fuel at less than 1 wt %.

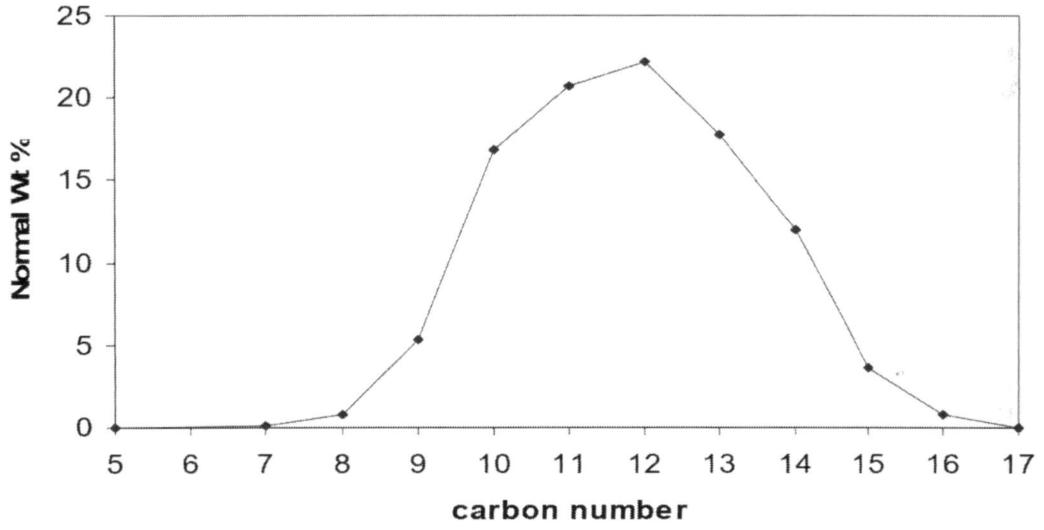

Figure 1: *Jet-A: probability density function with n values rounded to the nearest integer.*

Experimental Section

Chemicals

The feed aviation fuel, Jet-A, was obtained from the Columbia Regional Airport near Columbia, Missouri. The carrier gas used for the Carbon-Hydrogen-Nitrogen (CHN) analyzer was ultra-high purity oxygen with a purity of 99.994%. Industrial grade nitrogen gas with a purity of 99.998% was used for the CHN's pneumatics. Calibration standards for the CHN were durene, sucrose, and EDTA (ethylenediaminetetraacetic acid) from the Leco Corporation. The feed water for the experiments was distilled and deionized.

Apparatus

The supercritical water reformation experiments were carried out in a custom-designed experimental process unit, a schematic process flow diagram of which is shown in Figure 2. The unit is comprised of the five main components: reactant feed system, reactor proper, ceramic heaters and insulators, integrated heat exchanger, product sample collection system, and data acquisition and control system. The reactant feed system consists of two Eldex high-pressure reciprocating micro-metering pumps. Deionized water was fed through a model BBB-4 pump, while an AA-100-S model pump fed Jet-A fuel into the system. The balance used to measure the mass of fuel fed was manufactured by Arlyn Scales, model D620L, with a range of 0-32 kg and reading increments of 0.002 kg. The balance for water mass measurement was an Arlyn-610C with a range of 0-44 kg and reading increments of 0.01 kg. The reactor was a tubular pressure vessel constructed of Inconel 625 (Grade 1) with a pressure rating of 33.6 MPa at 923 K and 24.2 MPa at 983 K. Four zone ceramic heaters were used to heat the reactor. Sample ports on the system allowed for collection of various liquid and gaseous products.[1,6,7]

The analysis of the coke products was performed using a Leco CHN-600 elemental analyzer. The CHN utilizes a furnace which combusts the sample. Separate IR cells are used to detect the amounts of carbon dioxide and water produced. The results are linearized and adjusted according to changes in barometric pressure and chemical interferences. The results are displayed as mass percentages on the command console. The mass of the cokes was weighed using a Leco LB-80 scale with a range of 0-100 g and a reading increment of 0.1 mg.

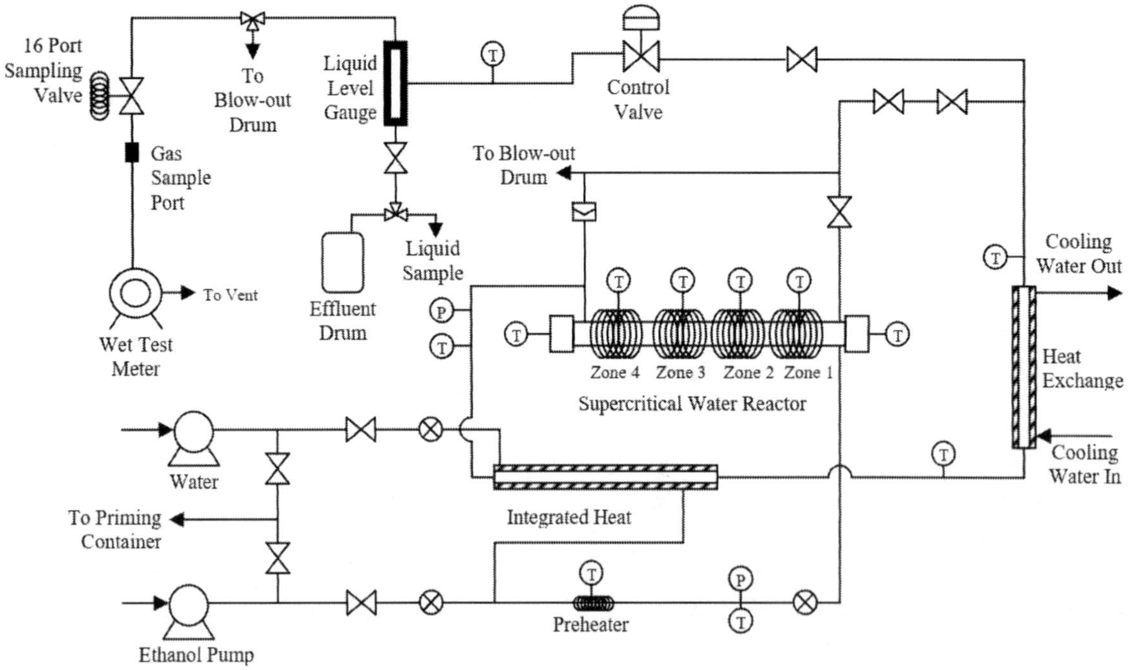

Figure 2: *Process flow diagram of supercritical water reformation unit at MS&T.*

Procedure

The reformation experiments were conducted in the following manner: system start-up to operating conditions, sample and data collection, and system shut-down. System start-up involved bringing the internal temperature and pressure of the supercritical water reactor (SWR) to operating conditions. Once the operating temperature was reached water was fed into the system at a desired flow rate and the temperature and pressure were allowed to stabilize before fuel was fed into the reactor. After stability of temperature, pressure, and reactant flow rates was achieved data acquisition began. The data acquisition system recorded inlet and outlet pressures, and temperatures at various locations along the system. As the experiments progressed, liquid and gaseous samples were taken for analysis. In addition, the masses of water and fuel fed into the system were recorded. When an experimental run concluded, operating conditions were changed to those for the next run and the system allowed to reach stability. When experiments were finished the system was shut-down and the reactor allowed to cool and depressurize.[6,7]

After the system cooled to ambient temperature, the reactor was removed from the system and secured on a table. The reactor was opened and a length of plastic tubing was used to remove the residual coke from the walls of the reactor. The coke was collected in boats constructed of aluminum foil placed at the ends of the reactor. The residues were transferred to plastic storage dishes and the mass of coke removed from the reactor recorded.

For analysis of the cokes, the CHN was zeroed with empty tin capsules. Blanks were combusted and entered into the data stacks of the CHN until the values read on the command console yielded values at or near zero. Approximately 0.6 g of standards of sucrose, durene, and EDTA with known percent compositions of carbon and hydrogen were placed in tin capsules and reacted. The results of the standards were entered into the CHN database and run until carbon and hydrogen values were obtained that correlated with the known percentages of the standards. Approximately 0.6 g of coke samples were placed in tin capsules and combusted in the CHN and the resulting percentages recorded.

Results and Discussion

A total of sixteen experiments were conducted in four sets with pressure effects being investigated above and below 22.06 MPa, the critical pressure of pure water. All of the experiments were conducted above 647 K, the critical temperature of pure water.[8] Table 1 shows four runs were conducted at subcritical conditions for pure water while twelve were conducted at supercritical conditions for pure water.

Table 1: Temperatures and Pressures of Subcritical and Supercritical Reformation Experiments Conducted

Set	Run ID	T (K)	P (MPa)
Subcritical			
1	WA_122	964	22.01
1	WA_123	973	20.65

1	WA_124	972	18.96
1	WA_125	969	17.24
Supercritical			
2	WA_126	986	24.01
2	WA_127	978	23.49
2	WA_128	958	23.70
2	WA_129	978	23.63
3	WA_113	969	23.87
3	WA_114	957	23.61
3	WA_115	974	23.97
3	WA_116	975	23.78
3	WA_117	981	24.05
4	WA_130	930	32.14
4	WA_131	927	32.54
4	WA_132	924	32.30

Coke residues were removed between each set of experiments yielding a total of four coke residue samples which were analyzed for percentage composition of carbon and hydrogen. Additionally, the total masses of Jet-A fed into the reactor was compared with the total mass of cokes removed from the reactor. The results are summarized in Table 2 and show that as the reactor pressure increased the amounts of carbon in the coke residues decreased while the amount of hydrogen in the coke residues increased. In addition, the ratio of fuel fed to coke produced increased with increasing reactor pressure.

Table 2: Effects of Pressure on the Ratio of Jet-A fed to Total Coke Removed and the Carbon and Hydrogen Contents Remaining in Coke

Set	$\overline{T}$	$\overline{P}$	Jet-A fed	Total coke	Ratio of Jet-A fed to Total coke	Carbon content in coke	H_2 content in coke
	(K)	(MPa)	(g)	(g)	(g_{fuel}/g_{coke})	Mass %	Mass %
1	970	19.72	1094	200.1	5.398	96.31%	0.9923%
2	975	23.71	935.0	134.8	6.937	96.28%	1.019%
3	971	23.85	682.0	51.60	13.28	96.26%	1.034%
4	927	32.33	1003	34.60	28.87	96.02%	1.280%

Figure 3 illustrates the cokes' composition with respect to carbon and shows a proportional decrease in carbon percentage with an increase in pressure. These findings indicate that more carbon products are leaving the reactor in liquid and gaseous phases and less carbon is remaining in the reactor as coke.

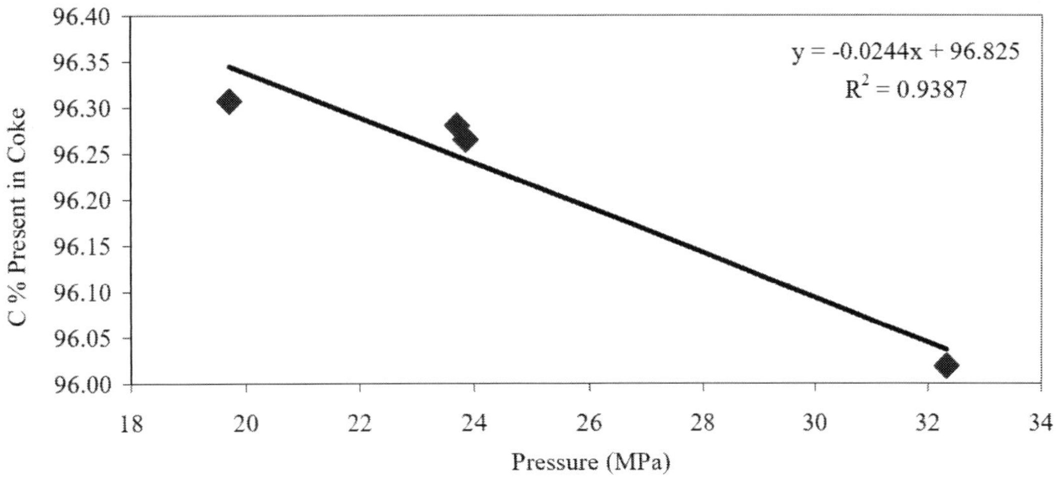

Figure 3: *Pressure versus carbon percentage remaining in coke.*

The trend of increasing hydrogen content in the coke with increasing reactor pressure shown in Figure 4 suggest that the hydrocarbons remaining in the high-pressure cokes contain more aliphatic chains in polycyclic structures and/or have lower polyaromaticity. This is important because it provides evidence that the polyaromaticity of the coke constituents are of a lower order with increased reactor pressure. Lower polyaromaticity in the coke will make the coke easier to remove from the reactor as well as prolong the soluble period of coke precursor species in the supercritical water reactor, thus contributing to the overall reformation reaction.

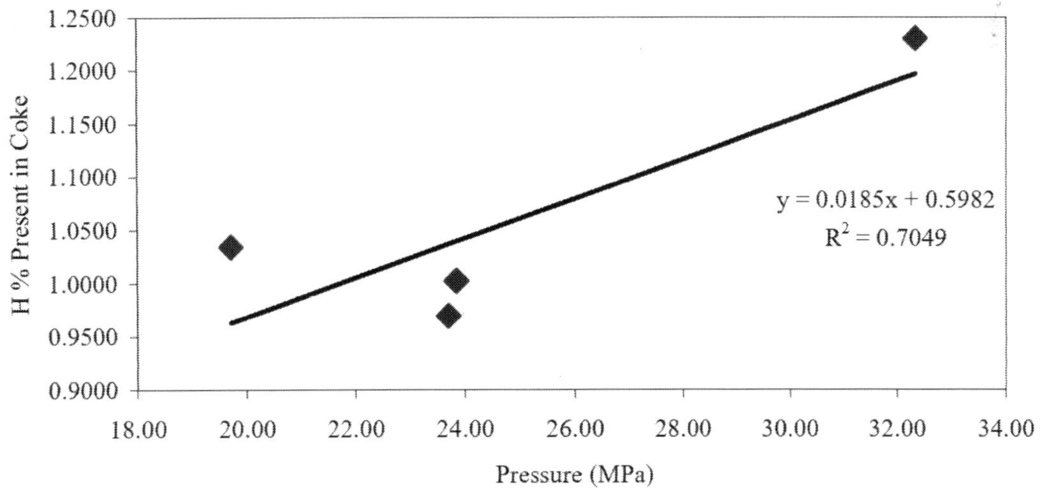

Figure 4: *Pressure versus hydrogen percentage remaining in coke.*

In Figure 5 the fuel fed to coke removed ratio increases with increasing reactor pressure. This finding shows by increasing the reactor pressure to values beyond the critical pressure of pure water more fuel can be reformed with a lower amount of coke being produced in the reactor. The significance of the supercriticality of the reactor conditions on coke formations is well manifested by this result.

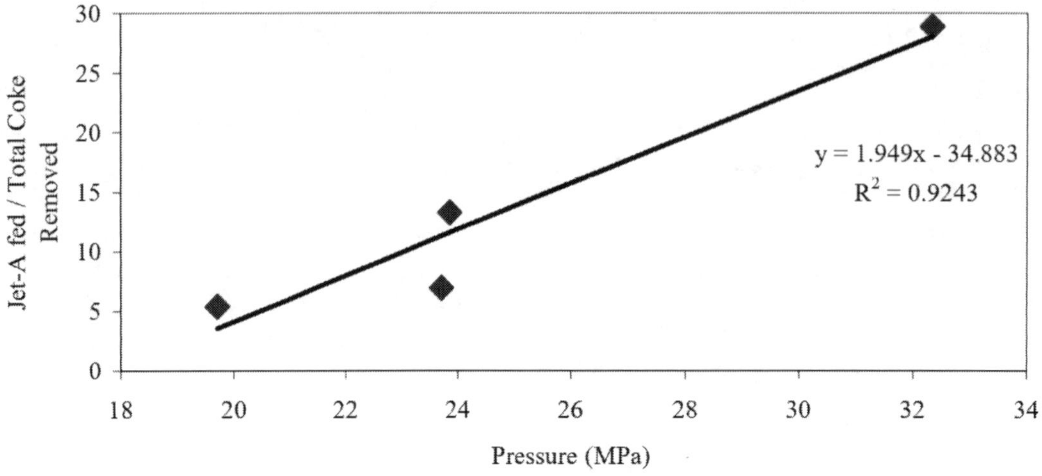

Figure 5: *Pressure versus ratio of Jet-A fed to total coke removed.*

Conclusions

A total of four coke residue samples produced as a result of Jet-A reformation to hydrogen were analyzed for carbon and hydrogen content utilizing a Leco CHN-600 elemental analyzer. Effects of pressure were determined above and below the critical pressure (22.06 MPa) of pure water on the amounts of carbon and hydrogen left in coke residues collected from the reactor. It was found as the pressure increased the carbon content of the cokes decreased and the hydrogen content of the cokes increased. This provides evidence that the polyaromaticity of the PAHs in the coke residues was lowered when using higher reactor pressures. The significance of this result is that PAHs with a polyaromaticity less than four are completely soluble in supercritical water giving the PAHs more time to reform before coking happens.[1,9] An additional finding as a result of this study was the total amount of coke produced was noticeably less with increased reactor pressure.

Acknowledgements

This work was fully sponsored by the U.S. Army under PE Number 0602705A through a subcontract from DRS Technical Services. The authors are grateful for insightful technological guidance and supervision provided by Dr. Terry Dubois, RDECOM, U.S. Army, Ft. Belvoir, Virginia.

References

1. Lee, S.; Lanterman, H.B.; Wenzel, J.E.; Edwards, N.; Adams, A.; Wootton, J.; Garcia, A. Noncatalytic direct reformation of JP-8 fuel in supercritical water. *Preprints-American Chemical Society, Division of Petroleum Chemistry* **2006**, *51*,

521-523.

2. Edwards, T. Harrison, W.E., III; Maurice, L.Q. In *Properties and Usage of Air Force Fuel: JP-8*, Proceedings of 39th AIAA Aerospace Sciences Meeting & Exhibit, Reno, NV, January 8-11, 2001; AIAA: Reston, VA, 2001; AIAA 2001-0498, 1-2.

3. *Jet-A Aviation Fuel* ; MSDS No. 1975 [Online]; Chevron Phillips Chemical Co. LP: The Woodlands , TX , July 31, 2001. www.cpchem.com/enu/JET_A_AVIATION_FUEL.asp (accessed Feb. 4, 2008).

4. ASTM Standard D2887, 2006a, "Standard Test Method for Boiling Range Distribution of Petroleum Fractions by Gas Chromatography," ASTM International, West Conshohocken, PA.

5. Lee, S.; Lanterman, H.B.; Wenzel, J.E.; Picou, J.; Factor, M.; Leavitt, L. Global kinetics of supercritical water reformation of JP-8 fuels. *Preprints-American Chemical Society, Division of Petroleum Chemistry* **2006**, *51*, 521-523.

6. Wenzel, J.E.; Niemoeller, A.; Stever, M.S. Kinetics of Reforming Ethanol into Hydrogen in a Supercritical Water Medium. Presented at *2007 AICHE Annual Meeting*, Salt Lake City, 2007.

7. Wenzel, J.E.; Picou, J.; Lee, S. Control of Ethanol Dehydration in the Supercritical Water Reforming of Ethanol into Hydrogen. Presented at *2008 AICHE Annual Meeting*, New Orleans, 2008.

8. International Association for the Properties of Water and Steam. *Guideline on the Use of Fundamental Physical Constants and Basic Constants of Water*; IAPWS, Gaithersburg, Maryland, September 2001, 5.

9. Brunner, E.; Thies, M.C.; Schneider, G.M. *J. Supercrit. Fluids* **2006**, 39, 160-173.

25a Rude Awakenings in Pilot Plant Operation

David C. Attride, Eastman Chemical Company, 200 S. Wilcox Dr., Kingsport, TN 37662

The pilot plant is a vital source of quantities of materials for testing and data for the design of commercial-scale equipment. Pilot plant operation often reveals subtleties in the process that were overlooked at the lab-scale. This paper discusses several pilot plant experiences where "simple" lab processes were difficult to operate at larger scale. One example describes a batch distillation which unexpectedly became a reactive distillation. Another example describes a continuous process with closed recycles that became virtually inoperable due to closure of the recycle loop. The pilot plant is an important step in answering many key knowledge gaps. It often contributes to the general but key knowledge gap "we don't know what we don't know."

73d Collaboration Leads to Successful Introduction of New Products

Eric H. Klingenberg and **Cawas A. Cooper**. Air Products and Chemicals, Inc., 7201 Hamilton Boulevard, R&D 3, Dock 6, Allentown, PA 18195-1501

This presentation will show what collaborative efforts were needed for taking concepts to commercialization of five new molecules for two growing businesses. Team of chemists, chemical engineers, and plant operators worked very closely in geographically separated laboratories and pilot plants for rapid scale-up. Roles played by each of the functional areas will be summarized to show elements that contributed to success and also emphasize pitfalls to avoid.

25d Planning An Ethylene-Based Complex in the Middle East

Osman Aboul-Nasr, Chemical Technology Center, Fluor, Greenville, SC 29607

This paper addresses the steps normally taken in developing and executing an ethylene-based complex in the Middle East. It examines the roles of the Local Owner, the Foreign Partner, the Technology Licensor and the Engineering Contractor/Consultant. It is important that the Local Owner involve the other parties in the right order, to improve the chances of success of the enterprise. This paper addresses the roles of the different players and the optimum sequence they enter the stage. Examples of successful projects will be presented.

25e Trends in Investment in Ethylene-Based Technology in the Middle East

Osman Aboul-Nasr and Lauren Hosking. Chemical Technology Center, Fluor, Greenville, SC 29607

The rise in the oil prices, growth in the petrochemical markets and the opening of the Middle East for foreign investor are the perfect condition for successful investment in the petrochemical sector. New industrial cities are planned either as an expansion of existing cities, or brand new cities. The latter, in some instances, is seen as a matter of national security.

This paper will address new ethylene-based complexes in the Middle East in various stages, from planning to execution, and their impact on the global petrochemical markets.

73c Herding Cats: Forming and Leading a Plant Troubleshooting Team

Jonathan Worstell, Hoder, Shell Global Solutions US, 3333 Highway 6 South, Room E-2061, Houston, TX 77082 and Richard E. Robertson, Geismar Manufacturing Facility, Shell Chemical Company, River Road, Geismar, LA 70734.

When plant performance declines, anxiety rises. The Business Group wants pounds and answers, in that order. Thus, the formation of a troubleshooting team. This presentation discusses the membership of such a team, how to lead such a team, and how that team interacts with the technical, operational, and safety organizations of the stricken plant.

206e Steam Traps: The Hidden Gold in Chemical Plants

Jonathan Worstell, Hoder, Shell Global Solutions US, 3333 Highway 6 South, Room E-2061, Houston, TX 77082

Steam traps are designed to separate condensate from steam. However, many simply vent steam, which is high dollar water vapor. This presentation discusses the theory of steam traps and their proper operation. It also discusses the cost of improper steam trap operation.

171e Distributed Reforming of Bio-Oil for Hydrogen Production

Jonathan R. Marda[1], Anthony M. Dean[1], Stefan Czernik[2], Robert J. Evans[3], Richard French[2], and Matthew Ratcliff[2]. (1) Chemical Engineering, Colorado School of Mines, Alderson Hall, 1600 Illinois Street, Golden, CO 80401, (2) National Renewable Energy Laboratory, Golden, CO 80401, (3) Hydrogen Technologies & Systems Center, National Renewable Energy Laboratory, Golden, CO 80401-3393

With the world's energy demands rapidly increasing, it is necessary to look to sources other than fossil fuels, preferably those that minimize greenhouse emissions. One such renewable source of energy is biomass, which has the added advantage of being a near-term source of hydrogen. Over 500 million tons of biomass could be available in the U.S. at less than $50/ton, which can be converted to 50 million tons of hydrogen [1]. The challenges for the variety of feedstocks that will be used include handling and drying, regional and seasonal availability and variability, and the impurities that could be generated and have an impact on conversion technology and hydrogen purity. Processes must be feedstock flexible, and minimize costs for feedstock collection, transport, and processing.

While there are several potential routes to produce hydrogen from biomass thermally, given the near-term technical barriers to hydrogen storage and delivery, distributed technologies such that hydrogen is produced at or near the point of use are attractive. One such route is to first produce bio-oil via fast pyrolysis of biomass close to its source to create a higher energy-density product, then ship this bio-oil to its point of use where it can be reformed to hydrogen and carbon dioxide. This route is especially well suited for smaller-scale reforming plants located at hydrogen distribution sites such as filling stations. There is also the potential for automated operation of the conversion system [2].

This research addresses the challenge of distributed hydrogen production with the target of producing hydrogen at $3.80/kg by 2012 [3]. A system has been developed for volatilizing bio-oil with manageable carbon deposits using ultrasonic atomization and by modifying bio-oil properties, such as viscosity, by blending or reacting bio-oil with methanol. Non-catalytic partial oxidation of bio-oil is then used to achieve significant conversion to CO with minimal aromatic hydrocarbon formation by keeping the temperature at 650°C or less and oxygen levels low. The non-catalytic reactions occur primarily in the gas phase. However, some nonvolatile components of bio-oil present as aerosols may react heterogeneously. The product gas is passed over a packed bed of precious metal catalyst where further reforming as well as water gas shift reactions are accomplished completing the conversion to hydrogen [4].

The approach described above requires significantly lower catalyst loadings than conventional catalytic steam reforming due to the significant conversion in the non-catalytic step. The goal is to reform and selectively oxidize the bio-oil and catalyze the water gas shift reaction without catalyzing methanation or oxidation of CO and H2, thus attaining equilibrium levels of H2, CO, H2O, and CO2 at the exit of the catalyst bed.

The results of experiments performed in a laboratory scale vertical reactor have shown that this process shows great promise as CO yields of over 50 % have been observed in the non-catalytic partial oxidation step and equilibrium yields of H2 (~80 %) have been observed after the catalytic step. The experimental results are being used to develop a process flow diagram (including mass and energy balance) for the process to determine whether the economic target discussed above is feasible. The results and conclusions of the efforts to develop this process will be discussed.

References

(1) Milbrandt A., A Geographic Perspective on the Current Biomass Resource Availability in the United States; National Renewable Energy Laboratory: Golden, CO, TP-560-39181, 2005.

(2) Czernik, S.; Elam, C.; Evans R.; and Milne, T., in Thermal and Chemical Biomass Conversion; Bridgwater, A.V. and Boocock D.G.B., eds.; CPL Press: Newbury, UK, 2006; pp.1752-1761.

(3) U.S. Department of Energy: Hydrogen, Fuel Cells and Infrastructure Technologies Program; Multi-Year Research, Development and Demonstration Plan, Section 3.1 Hydrogen Production; U.S. Department of Energy: Office of Energy Efficiency and Renewable Energy: Washington, D.C, 2006.

(4) Evans, R.J.; Czernik, S.; French, R.; and Marda, J.; Distributed Bio-Oil Reforming, DOE Hydrogen Program FY2006 Annual Progress Report, 2006.

Tar Wars in a Complex Chemical Plant: A Fundamentals Approach to Problem Solving

Larry G. Wright, Bruce M. Bell, J. Paul Chauvel, Chris P. Christenson, Bruce S. Holden, Terry McCabe, Ron B. Leng, Istvan Lengyel, Dana A. Livingston, Don B. Wood

Problem solving in a chemical plant requires using data and the fundamentals of chemistry and engineering to link "cause and effect" relationships between the chemistry, the process, the control system, and the problem. Once the underlying cause is identified, it is about resolving or minimizing the root cause(s) that can be manifest in a myriad of symptoms in the plant. That is the way problem solving should be approached. However, in manufacturing when the process goes awry, demands for pounds are high, financial losses mount, and frustration increases by the minute, it is easy to make changes to "respond" to the symptoms and "plant emotions" to show that "something" is being done. These reactions and short-cuts to "solutions", which are often not supported by data, typically fail to solve the underlying problem because they address the symptom. Although the problem may momentarily go away it will likely reappear in the future. In this case, more pounds and money is destined to be lost.

The case study used to showcase Dow's approach to problem solving focuses on how a decades-old fouling problem within a Dow plant was solved. The plant is part of a much larger chemical complex that produces several valuable products and intermediates for Dow. The complex is composed of highly integrated plants with multiple unit operations that use different organic raw material feedstocks. Streams from the individual plants also supply reclaimed feedstocks, catalysts, and chemical intermediates to each other within the complex. Outside the site this complex of plants supplies downstream intermediates for other Dow plants. As expected, asset utilization losses in any one of these plants quickly become upsets in all the plants when supplies are tight and market demands are high.

Figure 1. Photograph of the "plugged" "heavies" unit taken during a shutdown. The tars were observed to have the appearance of "gummy sands."

During the summer of '05, a high frequency of plugging incidents in a "heavies" unit (Figure 1 and Figure 2), used to separate catalyst and "heavies" from a lighter product stream, disrupted and limited sales for several of Dow's products. With the impacted plant only operational for periods of 3 to 9 days followed by plant shutdowns of 2 to 3 days for cleanup, each incident cost the business critical pounds in lost production for one of its primary products. Unfortunately, plugging was not a new occurrence for the plant, as the "heavies" unit had experienced severe plugging episodes in the past (Figure 2). Historically, with an average of 18 plugs per year, the annual cost to the business for this reoccurring problem was estimated to range between $3.5MM and $10MM (range tied to the product margin, product mix, and plugging frequency at the time). Several previous efforts in the plant had attempted to resolve the plugging problem over the years; however, those efforts were unsuccessful.

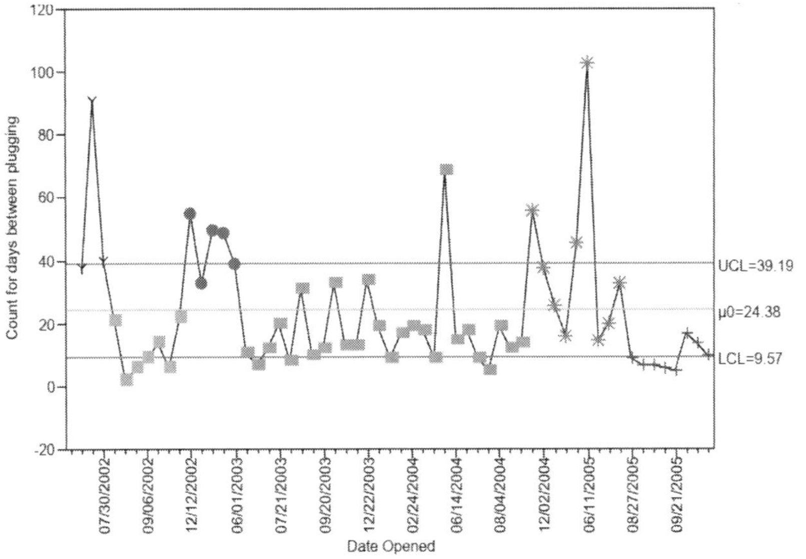

Figure 1. Control chart of "heavies" plugging events. The y- axis reflects the days between plugging and the x-axis is a record of the date at which plugging was registered.

Market demands and the earlier lack of success led the business to form a multi-discipline team of chemists and engineers to understand the chemical and engineering fundamentals of plugging. The objective for the team was to use this fundamental understanding to devise chemical, engineering, and control strategies to mitigate the effect of "heavies" so the plant could operate without plugging for a minimum of 5 months. This 5 month target defined "success" for the team. The team's investigative approach consisted of:

- characterizing the "heavies" as to composition, molecular structure, and thermal/mechanical stability,
- using characterization results to develop a chemical mechanism for "heavies" formation based on actual process conditions,
- developing analytical tools and sampling protocols to track tars and "heavies" precursors throughout the chemical processes involved, and
- implementing short-term operational changes in the process to test hypotheses concerning tars formation and stability.

Where possible, research was performed in parallel so that parallel processing of information could be utilized to further and expedite overall efforts. In all aspects of problem solving, hypotheses were formed, data was collected, and hypotheses were tested to drive decisions on future experimental paths and changes to the plant. Methodical experimentation and data-based decisions drove change and confidence in the plant operations staff.

This paper will discuss how that team worked together using the *weapons* of fundamental chemistry, engineering, the scientific method, and Six Sigma to *fight* the *Tar Wars.* The net result was that the team won the decades' old *battle* in less than three months by addressing the underlying cause of plugging in the "heavies" unit.

Tar Characterization and Mechanism for Tar Formation

Understanding the structure of the tar was required to develop a hypothesis as to how the tar was formed in the process. Solid tar material from the "heavies" unit was characterized by a combination of gas chromatography/mass spectrometry (GC/MS), liquid chromatography/mass spectrometry (LC/MS), liquid chromatography/mass spectrometry-mass spectrometry (LC/MS/MS), and gas chromatography /matrix isolation infrared spectroscopy (GC/IR). The tars were determined to be organometallic, with the inorganic part of the molecule coming from the catalyst used in the process. The tar ligands that were attached to the catalyst in the process are shown in Figure 3. The structures of the ligands were confirmed by isolating the ligands with preparative LC methods. The isolated materials were then characterized by C-13 nuclear magnetic resonance (NMR) and single crystal x-ray diffraction (XRD).

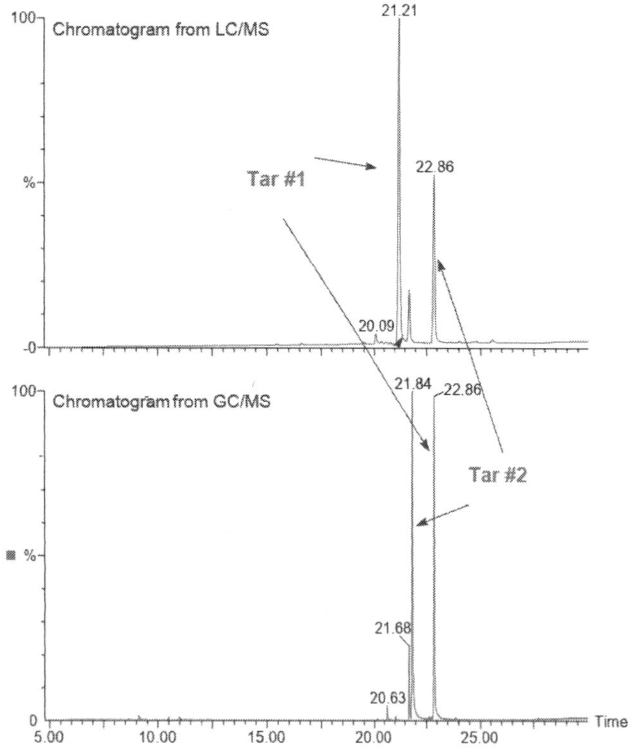

Figure 3. Electrospray LC/MS and GC/MS were used to separate the tars and determine their structures. Tar structures are not shown for proprietary reasons.

This characterization work identified the structure of the tars, led to the development of a chemical mechanism for their formation, and showed that oxygen and the catalyst was embodied in the two tar molecules. The mechanism, proven experimentally in the lab with water spiked process samples, supported the proposed mechanism and showed that the catalyst was a carrier of oxygen in the process.

Oxygen in any form, whether it be in the form of water, oxygen, or an organic or inorganic oxygenate, is an abnormal reactant in the process chemistries and could only be introduced through a combination of feedstock impurities, vacuum leaks, and/or wet maintenance procedures. The mechanism that was proposed showed that once oxygen and water entered the process, it reacted with the catalyst and the process feeds to produce tars (or tar precursors that led to tars) that plugged the "heavies" unit. Understanding the chemistry and the mechanism allowed the team to develop a grab-sample GC/MS method to track the buildup of tars and tar precursors in select process streams. The mechanism for tar formation and the ability to monitor tar precursors focused the team's efforts on identifying oxygen (water and air) incursion sources and removing those sources from the process. In this case, the hypothesis was that reducing oxygen in the process would reduce tar formation.

Oxygen Migration through the Process

Identifying the source of low-level oxygen incursion into a highly integrated process, whether it come in through the air (vacuum leak) or through a feedstock (water) as an impurity, is challenging. It is even more challenging in a process that has liquid, solids, organic, inorganic, and organometallic feeds, products, and reactants process. It is a difficult analytical problem. Neutron activation analysis (NAA) was used to measure total elemental oxygen in order to understand oxygen incursion, retention and migration throughout the process. The particular advantage of NAA for these samples was that it required minimal sample preparation (filling sample vials) and was amenable to the liquid, solid, organic and inorganic sample matrices encountered in the process. Although NAA did not characterize the specific forms of the oxygenated species in the process, it did provide coarse guidance for the team to identify gross changes in oxygen levels in specific pieces of process equipment.

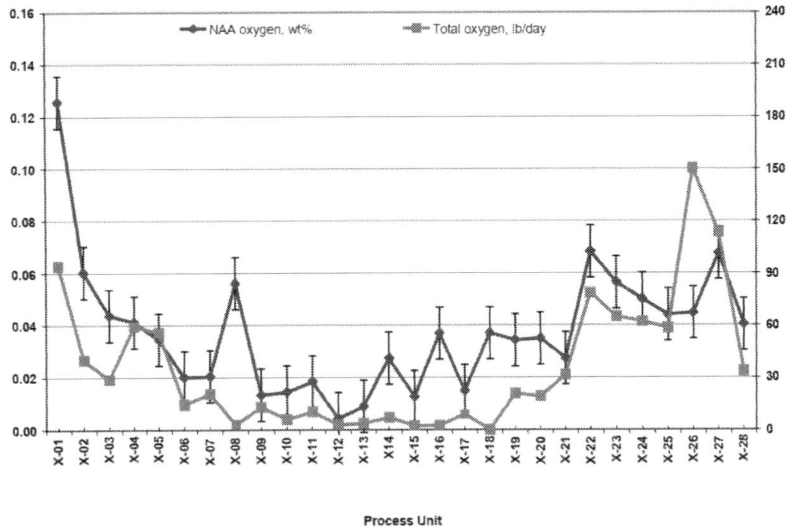

Figure 4. Snapshot of NAA data showing total oxygenates entrained in two plants. The left half of the plot (everything to the left and including X-17) shows the oxygenates in a front-end plant that feeds the back-end plant containing the "heavies" unit.

Figure 4 shows the information obtained using NAA results to perform an oxygen mass balance in the process. The plot shows two plants – everything to the left and including X-17 is the front-end plant, everything to the right of X-17 is the back-end plant that contains the "heavies" unit.

At a high level, the data showed that oxygen comes into the front-end plant through the feeds and was removed as lights through overhead streams to vents and/or as oxygenate impurities entrained in front-end product cuts as it migrates through the process. Here the "feed stream" for the plant that contains the "heavies" unit is being "cleansed of oxygen" as it passes from the front end to the back-end plant. However, it is not being completely cleansed as the data shows some oxygen being passed forward to the plant containing the "heavies" unit. Note also that the right side of the graph shows oxygen being "concentrated" in the back-end plant with the flow that moves into that plant.

Oxygen mass balance results generated from the NAA data indicated that three mechanisms allowed oxygen to enter the plant – cumulatively through the feed of the upstream plant, a distillation train, and through the reactor. The conclusion drawn from the mass balance work was that the "heavies" unit was consistently being feed a concentrated stream of oxygen. Although a highly oxygenated upstream stream is somewhat "oxygen purified" via various mechanisms as it moves toward the "heavies" unit, it still supplies the "heavies" train oxygen by using the catalyst as the carrier. The mass balance work also showed that the catalyst recycle stream "sinks and retains" the free oxygen that is introduced into the plant in various forms and releases that oxygen (oxygenates) back into the process via an equilibrium shift. Once oxygen comes into the "tar recycle sink", it can reside in the process for weeks participating in the various tar chemistries.

These results underscored the fact that limiting tar formation in the "heavies" unit was tied directly to keeping oxygen from entering the process and taking it out of the process once it had entered. These results also showed that few levers existed in the plant to alter the concentrations of oxygen that enters the "heavies" unit other than (i) reducing oxygen inflows from upstream distillation units, (ii) reducing the oxygen increase that occurs in an upstream reactor, and (iii) increasing the catalyst purge rate from the system in order to permanently remove oxygen from these processes. Options (i) and (ii) were evaluated by testing the feeds and the catalyst coming into those units for the presence of water, yet solutions were not available beyond what was currently being done in the plant to resolve the tar issue. The contribution of oxygen for these sources was also small. Hence, the catalyst purge solution was implemented to control tar formation.

The catalyst purge rate in the process was set based upon results from the GC/MS method to track tar precursors (identified as mechanistically import in the tar chemistry). Control limits for the tar precursors were held below the threshold limits that would cause the "heavies" unit to plug. Using the catalyst purge to take oxygen out of the plant was critical to controlling tar formation, as the recycle stream was determined mechanistically to be a "reversible oxygen sink" that allowed the recycled catalyst to accumulate and continuously supply oxygen to the process. This improvement (solution) that was implemented supported the hypothesis that the catalyst was a oxygen carrier and that the plugging could be reduced if oxygen was taken out of the process via increased catalyst purge.

Tar Thermal Stability

Flashing in the "heavies" unit occurs when feed material flashes across a control valve in the top head of the unit. The lower pressure downstream of the control value causes the liquid temperature to drop and the heavier components in the cooler liquid to concentrate. Tars at or near their solubility limit then precipitate and accumulate on the top tubesheet (Figure 1). While the tubes were designed to protrude above the tubesheet in order to provide a consistent liquid flow into each tube, in practice the design creates a sedimentary location for solids to collect. The thermal properties of the tars that form in the unit were examined by differential scanning calorimetry (DSC) and by hot-stage microscopy to determine if it was possible to shift tar solubility and mechanical stability within the process constraints of the unit. This was the hypothesis that was tested.

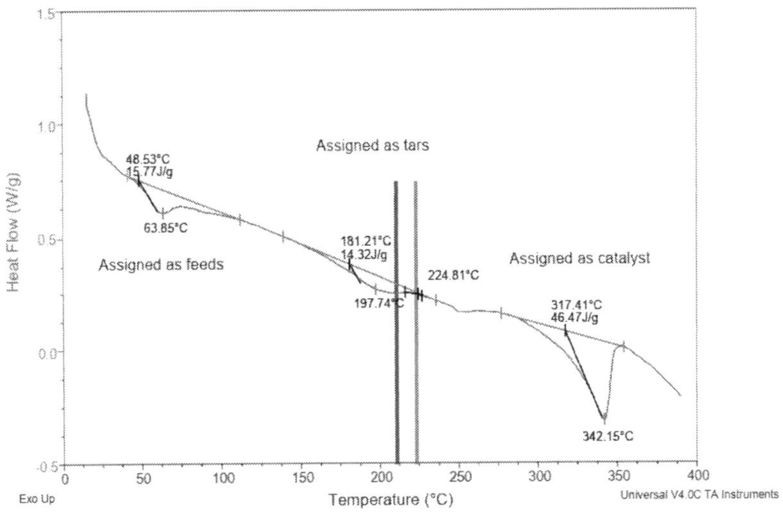

Figure 5. DSC of a tar showing an exotherm beginning at ca. 180 °C. The blue vertical line shows the temperature of the "heavies" unit when a high frequency of plugging occurred. The red vertical line shows the temperature at which the unit currently operates.

Figure 5 shows the DSC trace collected on the tar material. DSC showed a phase transition with the tar occurred at 200 °C. One of the more significant findings of the DSC experiments was that the tars did not exhibit the endotherm when they were heated a second time. This was attributed to the formation of a "structure" that melts around 200 °C that does not readily recrystalize.

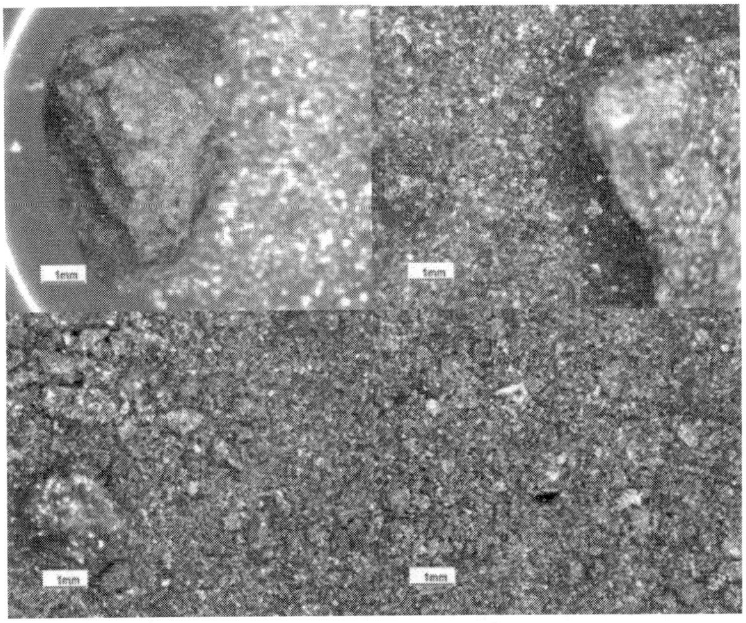

Figure 6. Photographs of tars exposed to different oven temperatures. Tar solids, sealed in headspace vials, were heated for 1 hour at 200°C (top left), 210°C (top right), 220°C (bottom left) and 230°C (bottom right) and then photographed.

The DSC observation led to oven and hot-stage microscopy studies of the tars to determine if it would be possible to make them mechanically unstable by raising the "heavies" unit temperature. If they did not agglomerate, it was hypothesized that they could be fluidized in the process. Hot-staged microscopy and the oven observations (Figure 6) validated the DSC results that a phase transition with the tars occurred around 200 °C. As a result of this study, the "heavies" unit temperature was raised by 12 °C (210 °C to 222 °C, refer to Figure 5) and the back pressure on the unit was raised. These changes would be expected to reduce precipitation and cause the process liquid that pools on the top tubesheet to increase and thereby reduce the tar buildup on the unit.

Conclusion

A methodical approach to problem solving is critical to implementing sustainable solutions in manufacturing. In all aspects of problem solving, it is imperative that hypothesis be formed, data collected, and the fundamentals of chemistry and engineering be used to link "cause and effect" relationships. Once the underlying cause is known, it is about developing sustainable solutions. In the case study presented here, fundamental knowledge about the tar's chemistry, its mechanism for formation, its solubility/mechanical stability, and the migration of oxygen through the plant process led to:

- increasing the tar purge to remove oxygen and thereby lower tar concentrations in the process,
- increasing the "heavies" unit back pressure and shell temperature by 12 °C to reduce mechanical stability of the remaining tars, and
- continuous monitoring of tars and tar precursors in the plant to maintain acceptable (non-plugging levels) levels.

The decades' old *Tar Wars' battle* was won in less than three months by addressing the underlying cause of plugging. The team's goal of 5 months was far exceeded, as the plant has experienced one year of "plug-free" running. The sustainable solution implemented in the plant today is supported by the photograph in Figure 7. Fundamentals won the *battle*!

Figure 7. Photograph of a spotless "heavies" unit after 81 days of continuous operation. The unit was opened for inspection during a normal shutdown to access the improvements. Only 4 tubes had tar build-up and that tar was glassy in appearance (unlike the granular tars observed in Figure 1).

25b Consequences of Pilot Plant Design Decisions

Adam Whalley, Zeton, Inc, 740 Oval Ct., Burlington, ON L7L 6A9, Canada

Often when designing a pilot plant, one of the engineer's biggest challenges is to design a plant that will meet objectives within a defined, or over constrained budget. One of the major pitfalls involves designing smaller pilot plant and demonstration plant unit operations using full scale principles. In most cases, equal or better operating results can be obtained while meeting budgetary constraints by using philosophies suited to the scale. The purpose of this presentation is to introduce professionals to philosophies that lead to successful pilot plant projects, while saving the company money. In addition, examples of decisions made during the design phase, and their associated positive and negative consequences during start up and operation, will be discussed.

Scale-down Considerations for Pilot Plants and Bench-Scale Units

Joseph B. Powell, Ph.D.
Chief Scientist, Chemical Engineering
Shell Global Solutions US Inc.
Westhollow Technology Center
Houston, Texas 77082

Overview

Correct sizing of pilot plants and bench-scale units is a critical decision in process development. Larger units allow closer approach to commercial-scale unit operations, but require more time and resources to implement, and present greater challenges in risk abatement and safeguarding. Smaller units may be deployed and modified more rapidly at lower cost, but unless properly designed, may not address key issues in design and scale up, or may not produce sufficient product to evaluate commercial success.

Decision factors for pilot plant scaling

Once a process concept passes the screening phase and a decision is made to pilot, a key question to ask is "what drives the scale of the pilot?" R&D spending increases dramatically at the pilot phase, and differences in pilot scale can impact costs by more than an order of magnitude, and may consume valuable resource which could be deployed on parallel developments.

Often, commercial interests drive the scale of a pilot program. For new product development, a threshold quantity of material may be required for product testing, or for production testing in a subsequent step. Required pilot plant scale can be simply determined from the project timeline and required product production rate. Since size is not driven by scale needed to demonstrate specific unit operations, the cost of operating at larger scale in "production mode" should be made clear to the business sponsor. While capital costs for pilots may scale at a much lower factor relative to recommended scaling factor exponents of 0.6 – 0.7 for commercial plants,[1] the difference in production scale between operation in production mode vs. minimum scale for demonstration of unit operations can in many cases be much larger than factors considered for commercial plants. I.e., a 20-fold increase in capacity (C) at an exponent of 0.5 typical of pilot plant pricing still results in a 5-fold increase in ultimate price P).[1]

$$P = P^{O} \times (C/C^{o})^{0.5}$$

Often, scale selection is driven by the *wow factor*, or a perception of how big the unit should be to impress a perceived, relatively non-technical investor or customer. Getting to the pilot plant phase for a pilot of impressive size can indeed become the ultimate goal of a development program if outside investment is targeted. This goal may lock designs and accelerate pilot construction at a pace beyond what is prudent from process development risk, and can lead to disappointment if the risk of acceleration is not communicated to funding sponsors, and ultimate commercial-scale costs of the piloted technology prove unacceptable.

To manage these risks, it is important for process development leadership to communicate openly with business sponsors. If test product production or technology showmanship are not the overall objective, then pilot-program goals are driven by prudent technology development and scale-up demonstration needs. A key strategy for pilot programs with the latter goal is to devise units to test key assumptions or potential show-stoppers as quickly and inexpensively as possible, while also allowing for future integration to address and demonstrate the remaining needs for a process design package, if successful. A smaller pilot may offer little advantage if a larger one is later needed to de-risk commercialization, except for cases where the number of process variables and options warrants multiple parallel units.

Minimum scale for unit operations

A key methodology for right-sizing pilot units whose scale is driven by technical considerations, is to rigorously ask the question "what is driving the scale?" Otherwise, one is likely to simply pick a size and scale that matches a previous development, without examination.

Is integration required?

The size of one unit operation can drive the overall pilot plant size, if full integration is attempted. A key decision in piloting is then whether complete integration is required, or whether certain larger-scale unit operations can be run in semi-continuous mode, using accumulated product from other steps in the process. Alternately, cold flow or mock-up models can be used to assess specific scale issues, with a smaller-scale unit incorporated into the pilot to perform the operation required to complete pilot integration, but not necessarily providing scale-up information. The ability to avoid full-scale continuous integration of all pilot steps in this manner is dependent on the particular scale-up risks identified.

Specific Unit Operations

Reactor scaling

Reactor scaling is typically dominated by transport phenomena, rather than chemical reaction.[2-3] Slow, direct reaction of liquid or gas streams is a rare objective. Most typically, reactor design entails challenges in fast homogenous reactions, or catalytic reactions where transport to or from an active site, may be limiting, and contacting of multiple phases (liquid-solid, gas-solid, gas-liquid) dominates the scaling problem. For fast liquid- or gas-phase reactions, mixing is key. Design of pilot apparatus for rapid liquid mixing can be treacherous, given inability to simultaneously scale power input per unit volume, and tip speed parameters.[4] Scaling rules for catalytic, multiphase reactors are dominated by the characteristic size of catalyst particles, or gas or liquid droplet formation. Characteristic dimensions for fixed-bed catalyst particles, liquid droplets or gas bubbles are typically on the order of 1 – 3 mm, such that many different reactor types ultimately result in similar minimum scales, where reactor diameter should approach (at minimum) 10 particle or droplet diameters, and length should approach 20 characteristic diameters to avoid entrance effects and minimize the impact of axial dispersion.[2] The fixed-bed limit in particle size is set by commercial pressure drop considerations (pilot scale largely unaffected), while droplet size is set by minimum allowed and maximum sustainable bubble or droplet size, set by interfacial tension. Similar scaling considerations for these diverse systems mean that a 25-mm or "1-inch" diameter reactor that is at least twice as long as wide, can address a wide range of reaction systems, from fixed-bed reactors with catalyst pellets, to multiphase gas-liquid reactors with finely divided slurry catalyst. In some cases, tricks may be played to improve transport at a smaller scale, e.g. via use of more finely divided solid diluent in trickle bed reactors to achieve 10:1 average particle to fluid diameter.[3,11] Care must be taken in scale down however to insure that maximum bubble diameter does not approach tube diameter, leading to a lab reactor flow regime (slug flow) that is not representative of commercial.[5]

If heat transfer is an important consideration for piloting, then the fact that heat transfer area per unit volume scales inversely with diameter is of considerable importance. Laboratory reactors lose heat due to this scaling effect, such that exothermic reactions are not apparent. Calorimetry or *a priori* calculation is required to assess heat release. For one recent example, an exothermic reaction with adiabatic temperature rise on the order of 70 °C was successfully diagnosed by careful measurement and calibration in a standard 300-ml stirred batch reactor, where the observed exotherm was only 2 - 3 °C. Temperature gradients of adiabatic commercial reactors must be self imposed via heat tracing, or commercial performance must be assessed by an appropriate nonisothermal reactor model using kinetics derived from the lab reactor via isothermal experiments at varying temperature.

For commercial reactors which must be designed for nonadiabatic conditions to optimize selectivity or prevent reaction runaways, the 25-mm or one-inch reactor remains a good choice relative to practical size allowed for a shell-and-tube heat exchanger reactor configuration (microchannel devices notwithstanding). The lab reactor now serves as a single-tube model of a commercial tube sheet, and uniformity of packing and flow distribution becomes the scale up challenge.

Separations unit operations

Thermal separations or "distillation" dominates the separations area, due to cost and simplicity. The same [surface area / volume] $\propto$ [1/diameter] scaling rule is also problematic however for distillation pilot plant equipment. Heat losses can be prolific, especially for metal columns, including those at 1-inch diameter scale. If compositions and physical properties are known, then temperature compensation can be applied, such that a process simulator can be used to predict the correct profile. If not, and if a large number of stages are needed for separation, then a larger diameter vacuum-jacketed glass column may be optimal (50 – 75 mm diameter), to minimize impact of heat losses.

Lab distillation apparatus is generally designed to give shorter characteristic distances to obtain vapor-liquid equilibrium, or smaller HETP (height equivalent to theoretical plate) on the order of 1 – 4 inches.[7,8] Oldershaw perforated plate columns may be employed with a rule of thumb of 0.5 tray efficiency, while use of protruded metal or gauze packings can provide for low HETP with use of correlation to relate back to commercial size column internals.[6-8] If a primary challenge for the pilot program involves distillation separation efficiency of a fine distillation, then commercial-scale packings may be required with HETP on the order of 0.5 – 2 feet.

Attempting to match commercial-scale temperature programs in jacketed columns, or especially commercial-scale HETP using commercial column internals, can give rise to large and expensive pilots, if directly coupled in real time. Often however, precise study of distillation performance at the pilot scale is not necessary. Batch distillations as small as 100-ml at infinite reflux with marker compounds of known relative volatility can be used to calibrate physical property data decks in process simulators. This approach has indeed been successfully used with correlations for commercial packings, to design 100-foot 80-stage commercial columns. Use of lab-scale internals to reduce HETP is a substantial cost and scale saver, if appropriate tie points can be made to performance of commercial-scale column internals. In some cases, it may not be necessary to run a fully integrated pilot plant with real time distillation recycle, in order to establish process viability. Collection of intermediate product to feed to a larger-scaled distillation train operated in campaign mode, with recycle of byproducts and solvents as appropriate, can often be used. Appropriate sampling and analysis can then allow calculation of steady state compositions, which can subsequently be tested via direct preparation of model compositions. In this manner, a difficult separation requiring a large distillation column or train does not necessarily have to set the size of the remainder of the pilot plant, especially where process simulator performance can be calibrated with high confidence via off line experiments and tie points.

Similar issues apply for separation steps such as fixed-bed adsorption or ion exchange, or liquid extraction.[10] Fixed-bed systems are subject to the same particle-to-tube diameter (d_p/d_t) and particle to length (d_p/L) constraints as packed bed reactors.[9] However, the performance can often be modeled from small scale (2 – 10 ml) batch measurement of adsorption isotherms combined with literature correlations for mass transfer. Given extensive correlation database, it is generally not necessary to establish the viability of the unit operation within the pilot unit itself; a smaller particle bed offering improved mass transfer can be employed to mimic commercial performance, if needed. Similarly, packed extractors using lab-scale internals analogous to those employed in distillation, or mixer-settlers may be employed for liquid extraction in order to achieve separations and compositions required for an integrated pilot. Limited off-line testing in larger units can be used to test scale up as needed.

In summary, pilot plant sizing can be minimized by deciding which are the key unit operations which must be conducted at the proper scale within the integrated pilot, in order to derisk commercial scale up. Other unit operations may then be conducted with nonrepresentative equipment or internals in order to produce the desired compositions and separations for the integrated pilot, if scale up of these operations can be done from first principles, or via data from limited off-line testing. The strategy of directly validating only key unit operations (vs. requiring direct continuous validation of all steps) can lead to smaller, more adaptable pilots where resource constraints or ability to test alternate configurations is a prime consideration. Cost and resource savings via this approach can be more than 10-fold.

References

1. M. S. Peters and K. D. Timmerhaus, *Plant Design Economics for Chemical Engineers*, McGraw-Hill, NY (1991).

2. D. E. Mears, *Tests for Transport Limitations in Experimental Catalytic Reactors*, Ind. Eng. Chem., Process Des. Dev., *10*, (1971), 541-547.

3. Koros, R. M. Scale-up Considerations for Mixed Phase Catalytic Reactors. *In Multiphase Chemical Reactors;* Rodrigues, A. E.,Calo, J. M., Sweed, N. H., Eds.; NATO Advanced Study Institute Series E; Sitjhoff & Noordhoff: Aalphen aan den Rijn, The Netherlands, 1981; Vol. 52, p 429.

4. N. Harnby, M. F. Edwards, A. W. Niewnow, *Mixing in the Process Industries*, Butterworth-Heinemann, London (1992)

5. Wen-ching Yang, *Handbook of Fluidization and Fluid-Particle Systems, CRC Press, 2003.*

6. J. R. Fair, H. R. Null, and W. L. Bolles, *Scale-up of Plate Efficiency from Laboratory Oldershaw Data*, Ind. Eng. Chem. Process Des. Dev., 22, 53-58 (1983).

7. M. R. Cannon, *A New Distillation Packing*, Ind. and Eng. Chem., 41, 1953-55 (1949).

8. J. R. Hufton, J. L. Bravo, J. R. Fair, *Scale-up of Laboratory Data for Distillation Columns Containing Corrugated Metal-Type Structured Packing,* Ind. Eng. Chem. Res., 27, 2096-2100 (1988).

9. M. D. LeVan and T. Vermeulen, *Channeling and Bed-Diameter Effects in Fixed-Bed Adsorber Performance*, AIChE Symposium Series, 233, 34-43 (1984).

10. R. E. Treybal, *Mass Transfer Operations,* McGraw-Hill, NY (1987).

11. S. K. Bej, R. P. Dabral, P. C. Gupta, K. K. Mittal, G. S. Sen, V. K. Kapoor, and A. K. Dailai, *Studies on the Performance of Microscale Trickle Bed Reactor Using Different Sizes of Diluent*, Energy and Fuels, 14, 701-705 (2000).

171b Microchannel Process Technology Unlocking Offshore GTL Opportunities

Jeff McDaniel, Daniel Weidert, David Kilanowski, and Robert Litt. Business Development, Velocys, Inc., 7950 Corporate Boulevard, Plain City, OH 43064

Natural gas without access to the world market, known as associated gas, is both a challenge and an opportunity. Due to ever stricter flaring regulations, associated natural gas is typically seen as a cost of oil exploration and production because of the capital equipment and energy required to re-inject the gas back into the reservoir. The cost of associated gas is highest offshore, where drilling wells and installing equipment are more expensive, and deck space is at a premium. The opportunity lies in the ability to monetize the natural gas through a Fischer-Tropsch gas-to-liquids (GTL) process enabled by microchannel technology; thereby, transforming the burden of stranded gas into a valuable resource that increases revenues and stretches reserves. The GTL application is a tremendous opportunity as over 150 billion cubic meters of natural gas are flared annually, while a substantial additional quantity is re-injected. The flared natural gas represents 5.5% of the world's annual production, and would be valued at $40 billion if it were delivered to the market. This level of flaring also needlessly adds approximately 375 million tonnes of CO2 per year to the atmosphere.

The vision of utilizing offshore natural gas to produce high value fuels has been around for years, but a number of factors, including the limitations of conventional processing hardware operating in the offshore environment, have prevented implementation. The primary challenges of operating a GTL process offshore are: 1) the ability to withstand high-intensity wave motion, 2) equipment size and weight, and 3) freshwater consumption. The wave motion experienced by floating production, storage and offloading (FPSO) vessels can cause havoc with process equipment, especially vessels with a high center of gravity or large liquid volume. Modular, low profile microchannel technology steam methane reforming (SMR) and Fischer-Tropsch reactor units can alleviate this concern by performing reactions rapidly in thousands of parallel channels; thus, minimizing reactor inventory and susceptibility to wave moments.

The compact nature and other aspects of microchannel technology make this approach especially attractive for offshore operations. The process intensification possible with microchannel process technology improves volumetric productivity and efficiency, reducing capital cost and shrinking facility footprints. Also, the low process water requirement of microchannel-based process flow sheet minimizes the need for desalination. All of these advantages result in a more cost effective offshore synthetic fuels process.

Sustained high oil prices, concerns about 'peak oil', and the push for cleaner transportation fuels have intensified interest in alternative fuels, including those made with Fischer-Tropsch based synthetic fuel processes. To address this global trend, Velocys, Toyo Engineering and MODEC are developing a microchannel based synthetic fuels process specifically for offshore applications. This presentation will discuss the status of the technology and specifics about commercial FPSO installations.

www.velocys.com

171c Experiences in Thermochemical Biomass Conversion

Calvin Feik, Daniel Carpenter, Steve Deutch, Katherine Gaston, and Steven Phillips. National Bioenergy Center, National Renewable Energy Laboratory, 1617 Cole Blvd, Golden, CO 80401

The Thermochemical Process Development Unit (TCPDU) at the National Renewable Energy Laboratory (NREL) is a half-ton per day pilot plant that can be used to study and collect data on a variety of biomass feed-stocks and is capable of producing a variety of products ranging from pyrolysis oils to syngas to hydrogen-rich gas. Process gas slipstreams can be diverted to various smaller reactor systems to evaluate fuel synthesis and gas clean-up methods.

Since the start of operations in 1996, the TCPDU has evolved mechanically (biomass feed system, char collection), chemically (catalytic reforming, gas scrubbing), analytically (MBMS tar measurement, GC heteroatom measurement), and in terms of process control and automation.

The lessons-learned over the last eleven years and the current capabilities of the TCPDU will be discussed.

Web Page: www.nrel.gov/biomass/thermochemical_conversion.html

233e Fast and Precise Quantification of Polymer and Petrochemical Products and Processes Using Automated TD-NMR Technology

Supriyo Ghosh, minispec Division, Bruker Optics Inc., 2700 N Crescent Ridge Dr, The Woodlands, TX 77381

The upstream and downstream processes in petrochemical and polymer industries involve application of various unit operations on raw materials. These processing steps induce multiple physico-chemical changes leading to the desirable properties in the final product. During the process development as well as the industrial-scale implementation of the same, monitoring of the effects of each processing step need to be fast, accurate and precise. TD-NMR (time domain nuclear magnetic resonance) is a technique which can be utilized to look into the products and physicochemical processes at a molecular level and precisely quantify them, using a small bench-top instrument. It obtains the signal from the atomic nucleus in a material. Mathematical analysis of the TD-NMR signal reveals the changes occurring in molecular level in a material. Measurements can be done in a matter of a few seconds to obtain dynamic, quantitative characterization of materials. The entire measurement procedure can be done in a continuous fashion using automated robotic system. In this presentation, the basic principle of TD-NMR technology will be introduced. Various high throughput automated TD-NMR characterizations relevant to product characterization and process development for petrochemical and polymer industries will be elaborated including but not limited to: extent of polymerization, crosslink density, crystallization, copolymer ratio, polymer density, finish on fiber, hydration, hydrogen content in fuels, impurity quantification, drying kinetics, solid-liquid ratio, emulsion stability, droplet size distribution, porous media microstructure and transport. Probable new areas of in the same industry where this TD-NMR technology can be exploited in the future will also be discussed.

AUTHOR INDEX